C0-AYN-565

The Modernization of the Chinese Salt Administration, 1900–1920

Harvard East Asian Series 53

The East Asian Research Center at Harvard University administers research projects designed to further scholarly understanding of China, Japan, Korea, Vietnam, and adjacent areas.

Sir Richard Dane

The Modernization of the Chinese Salt Administration, 1900–1920

S. A. M. Adshead

Harvard University Press
Cambridge, Massachusetts
1970

Distributed in Great Britain by Oxford University Press, London

Preparation of this volume has been aided by a grant from the Ford Foundation

Library of Congress Catalog Card Number 77–120315

SBN 674–58060–5

Printed in the United States of America

In Memory of
James Miller Vine

Carissimus Socer

Acknowledgments

I should like to thank first and most of all Professor John K. Fairbank, who supervised this work in its dissertation stage and made many suggestions for its improvement and development. I am also indebted to Dr. Thomas A. Metzger for his expert guidance in the history of the Chinese salt administration and to the late Esson M. Gale through whom I obtained basic source materials for the foreign gabelle. I am grateful to Tseng Yang-feng, formerly of the Chinese Salt Inspectorate, for giving me his personal reminiscences of Sir Richard Dane; to Dr. P. O'Brien for valuable advice; and to Professor Michel Mollat for no less valuable bibliographical material.

I must acknowledge the assistance of several eminent scholars and research institutions. In particular, I wish to thank Professor Kuo T'ing-i of Academia Sinica; Dr. Edwin G. Beal of the Library of Congress; Dr. Ts'ao Ching-ch'üan of the Salt Directorate, Tainan; Mrs. Mutsuko Yabu of the Toyo Bunko; and the Union Research Institute, Hong Kong. Transcripts of Crown copyright records in the Public Record Office, London, appear by permission of the Controller of H.M. Stationery Office. The frontispiece, reproduced by permission of the Hutchinson Publishing Group, is from *The Foreigner in China* by O. M. Green.

Finally, I thank Miss D. M. Dane, Sir Richard Dane's niece, for her correspondence with me. I hope that this book will contribute to the archives of a distinguished family of colonial administrators.

SAMA

Christchurch, New Zealand

Contents

Tables

The Modernization of the Chinese Salt Administration, 1900–1920

1 | The Structure of the Salt Administration in 1900

Research into nineteenth- and twentieth-century China has centered on the process described as modernization and, in particular, on the relative parts played in it by the traditional society and by the challenge and example of the West. This study will examine the modernizing process in a single institution, the salt administration, where between 1900 and 1920 a real if limited modernization was carried through by late Ch'ing officials and reformers of the early republic on the one hand, and by foreign administrators and advisers on the other. An analysis of the changes attempted or introduced may cast some light on the general problem of modernization in China.

In all traditional societies, salt, Pliny's *necessarium elementum,* was a key commodity, intimately related to politics, fiscality, and economic structure, "commerce essentiel, omniprésent, éveillant très tôt et de façon définitive les intérêts des princes." In China, in particular, topography and political tradition combined to give to salt a unique importance as the most accessible source of income and, after land, the most consistently taxed. The *Yüan-shih* commented: "Of the nation's assets, that which brings in the greatest profits is certainly salt." Unlike the ginseng monopoly, salt was a mass commodity in daily use, and, unlike timber or tea, the various stages of the trade were the object of daily interference by the government. A prominent and integral part of the imperial sys-

Note: Throughout the text and the notes all figures quoted in dollars are understood to be Chinese currency unless otherwise indicated. During the period under study the Chinese dollar was worth U.S. $0.50.

tem, the reconstruction of the salt administration is, therefore, a good subject for a study of modernization in China.[1]

Both an ostensive and a formal definition of modernization will be presented in the course of the argument, and these will be brought together in Chapter 7. The concrete definition may serve to familiarize the reader with the general historical background, while the abstract formula provides an analytical framework within which to describe the changes taking place.

Ostensively, then, modern society may be said to be separated from the traditional by technological, economic, social, and political differences. A modern society is characterized by the use of powered machinery and new sources of power; by a high level of consumer spending and the elimination of famine and fatal epidemic; by a marked degree of social equality based on personal achievement rather than inherited status; and by mass politics, whether democratic or totalitarian. A traditional society, on the other hand, is characterized by the absence of machinery, or by reliance on eotechnic sources of power; by an economy of scarcity, in which insecurity of food supply, health, and property must be accepted; by social inequality and social immobilities; and by restriction of political participation to an elite, frequently a self-perpetuating elite. Modernization may be defined as the transition from one kind of society to the other.

In China, this process took place in the nineteenth and twentieth centuries, accelerating significantly between 1895 and 1925 with the building of the railways, the fall of the empire, and the repudiation of Confucianism. More particularly, the first two decades of the twentieth century saw a revolution in social and administrative institutions, organizations below the level of national politics but above that of elementary economic units — shops, farms, and old-style trade guilds. Sir John Jordan, the British minister in Peking, described this revolution in his report for 1910: "changes one sees on all sides, not only in the capital, but in many provincial towns. I am not here referring only to constitutional changes, but to such improvements as are apparent to any foreign observer — police, sanitation, street lighting, prisons, soldiers." [2]

In the Hankow area, for example, the number of branch post offices rose from 40 in 1910 to 134 in 1920. News media multiplied:

"To-day hardly a town of any importance is without its newspaper, and it is estimated that there are over 4000 dailies now in existence in China," wrote the Shanghai customs commissioner in 1921. Public health became accepted as part of the agenda of government: the North Manchurian plague-prevention service was initiated in 1912, and the first government mental hospital was opened in Peking in that year. In Peking also, Sidney D. Gamble noted in his survey, "those who live in the city are constantly amazed at the extent and efficiency of the work" done by the new police board, whose activities included not only public order but also fire prevention, street cleaning and lighting, and the operation of the two public hospitals. The establishment of modern Chinese banks, Shen Chia-pen's legal reforms, the Ziccawei typhoon-warning service, and the establishment of new kinds of schools (for girls, for the blind, for special techniques) are all examples of this revolution.[3]

Social and administrative institutions such as these, it may be argued, play a special role in modernization, because it is through them that people grow familiar with a modernized context of living. Novelty at the national political level is frequently remote, while technological change may be encapsulated within an industry and appear no less remote. Change at the intermediate level, on the other hand, has a wide educative effect, raising expectations of efficiency and satisfaction, and, by its own employment requirements, forming a reservoir of people with modern attitudes on which other institutions can draw. In this way, changes are proliferated, and the transition is made from sporadic examples of modern institutions to a totally modern society. The revolution in the salt administration thus formed part of a wider, contemporary movement of critical importance in the modernization process.

It may now be asked what this overall transition meant to the salt administration in China: what structural alterations it implied, who were the principal agents, and, in particular, what were the respective roles of Chinese and foreigners throughout. The terminal dates, 1900 and 1920, are implied in these questions. Prior to 1900, change in the salt administration remained well within the tradition, and foreign influence was marginal. 1901 saw the empress-dowager initiate a wave of general reforming activity on the

Chinese side which deeply affected the salt administration. It also saw the compilation of the Maritime Customs' massive report, *Salt: Production and Taxation,* which marks the beginning, on the part of Europeans, of increased concern with Chinese salt affairs, a concern which reached its climax with the establishment of the foreign inspectorate under the reorganization loan agreement in 1913. 1920, on the other hand, the terminal year, may be taken to mark the end both of the traditional bureaucracy and of the age of the treaty system in China. The May 4 movement saw the rejection of Confucianism as the basis for the Chinese state, while European statesmen, like George Nathaniel Curzon, had become aware that the treaty system could not long survive in the post-war political atmosphere.

Even between 1900 and 1920, however, modernization was not continuous. Considerable efforts at reform were made both by the officials of the last decade of the empire, Tsai-tse and Chao Erh-hsün for example, and by the revolutionaries of 1911 such as Teng Hsiao-k'o and Ching Pen-po; but the most significant institutional changes were effected between 1913 and 1918, when Sir Richard Dane, first foreign chief inspector, was effective head of the whole salt administration. Dane's work — the architectonics of his reform, its application in the different regions of China, and its reception by the Chinese salt interest — forms the core of the discussion.

Contemporaries took a high view of Dane's achievement. Chinese compared his impact to that of Gordon and Hart, and the British minister, Sir John Jordan, writing to Sir Walter Langley, under-secretary at the Foreign Office, argued: "It is not too much to say that the success of the Salt Administration has simply saved China and incidentally raised our prestige as a race of administrators far beyond even what Sir R. Hart had attained in the past. In this its second year, the Salt Revenue will probably exceed the Customs, and the two together are, let it never be forgotten, the mainstay on which the present régime depends for its existence." [4]

We, however, should be more ready to doubt the extent to which a retired Indian civil servant, who knew no Chinese, could understand, still less radically reform, an ancient Chinese institution, which Dane himself described as "enormously complicated and absolutely diverse in every district."[5] How much did Dane achieve? Did his reforms mark any decisive advance on the work of the pre-1913 reformers or on the program of the Chinese salt experts who were his associates from 1913 on? To answer these questions requires an analysis of the institutions of the gabelle and an evaluation of the changes introduced between 1900 and 1920.

China's salt monopoly had, in the first place, a purely technological aspect: the gabelle manufactured and distributed salt within the technological and market constraints of the day. Further, there was a commercial aspect. Here the dominating factor was reliance on inland water communications which both dictated the pattern of the trade and, by facilitating bureaucratic control at all stages, made the trade in effect a partnership between officials and merchants. The primary purpose of official control, however, was taxation, and thus the salt administration had a third aspect: its role as a fiscal machine. The salt administration can also be viewed, apart from its fiscal functions, simply as a bureaucratic system, which can be analyzed in terms of staffing, costs, degree of specialization, and so forth. Finally, the salt also had an ideological aspect: salt affairs were a recognized part of Chinese political science, and its values and assumptions affected the running of the monopoly at all levels. This analytical framework will be used both for an initial survey and for the final summing-up.

In 1900, the Chinese salt administration was in each aspect premodern in a precise sense to be defined below. By 1920, it will be argued, thorough modernization (again to be defined) had taken place in the fiscal and administrative aspects, and to a lesser extent in the commercial aspect as well. In these fields, Dane had effected a radical change, moving the institution beyond the limits of the traditional. This modernization, nevertheless, remained incomplete: the technology of salt, the ways in which it was produced and consumed, remained largely unchanged, while the commercial aspect retained a number of traditional features. On balance, however, a decisive transformation had taken place in the structure of the institution: a threshold had been crossed.

Technology: Alimentary Demand, Eotechnic Production

The technological aspect of the salt monopoly was shaped by physical rather than by social or political factors. It developed from the interaction of human biological needs with the material resources available in China and the general level of Chinese technology. What was premodern about it was common to Chinese life as a whole at that date: it was static, or, in Fei Hsiao-t'ung's expression, earthbound, in the sense of serving immediate ends, using readily available resources, and being geared to the rhythms of nature rather than of trade. This may be illustrated by the two chief components of the salt industry, the demand for salt and its supply.

The demand for salt. The demand for salt in China under the traditional gabelle came from those who required it for alimentary rather than for industrial purposes. Salt was delivered for final rather than intermediate consumption, a pattern which in the salt industry may be regarded as typically premodern. A modern industrial country such as the United States "requires several times as much salt for industrial purposes as is consumed in food," primarily for the production of soda, an essential raw material in the glass, textile, and soap industries among many others. The extent to which a salt industry is integrated with the alkali branch of the chemical industry is a measure of its modernization. In Europe, this began with the first factory for the synthesis of alkali from salt set up by Nicolas Leblanc in Paris in 1791.[6]

In China, however, in 1900, salt was still overwhelmingly a foodstuff rather than a chemical raw material. The first accurate statistics of the percentages of salt used for different purposes demonstrate this. Even in 1921 nearly 32 million piculs of salt were used in China for alimentary purposes, while on the industrial side only a million piculs were used for fish packing and 61,000 piculs were used for other industrial purposes. Not all alimentary salt was used as table salt by individual householders, much being used by the food-preparation industries. The customs commissioner at Newchwang remarked that in north China "Salted vegetables are eaten in enormous quantities; salted fish, shrimps, crabs, beef, and pork, and the favourite pickled beancurd all form articles of diet in universal use," while F. A. Cleveland, foreign chief inspector, 1930–1934, noted that "a large part of the salt seasoning in food is in the form of sauces, salted fish and meats, and salted vegetables." Vegetables in particular seem to have played the dominant role in the demand for salt by the food-preparation industries, which in medieval and renaissance Europe was played by the fisheries and the butter and cheese industries. In China, fishery salt was of importance only in peripheral areas such as the Chusan archipelago and Hainan, while dairy salt was a negligible factor except in Mongolia and parts of Manchuria.[7]

Various consequences for the structure of the trade stemmed from this emphasis on salt as a foodstuff. First, in these circumstances, the principal market for it was naturally to be found in the main centers of population. Half the 30 million piculs of salt produced in China were consumed in five relatively urbanized areas: the Yangtze delta, the Peking–Tientsin region, the West River delta (2 million piculs each), the central Szechwan basin (3 million piculs), and the more dispersed central Chinese market of Honan, Hupei, and Hunan (6 million piculs).[8]

Economically, the salt business remained what W. W. Rostow has called a "derived-growth" rather than a "supplementary-growth" industry,[9] that is, its long-term growth depended not on the increased demand from specific industries such as fisheries or the chemical industry, but on a general increased consumption of foodstuffs through higher population. Given the slow rate of population increase in China after 1850, such growth was comparatively slow and unspectacular, and the Chinese salt trade was therefore less dynamic and less responsive to demand than other salt trades which were oriented to the modern world of industry.

Short-run fluctuations were linked to the harvest rather than to the business cycle. This feature is well documented in the statistics of the 1901 customs report. The salt likin collector at Ta-t'ung, the chief importing center for Anhwei, noted the seasonal fluctuations: "In the 7th moon, it being then the season for manufacturing soy, the chief ingredient of which is salt, the demand again becomes brisk, while in the 9th and 10th months the sale reaches its highest point, the salt being much in demand at the end of autumn for the preparation of winter food." A nineteenth-century

Chinese salt expert, Wang Shao-chi, had already remarked that in Shantung "consumption of salt depended not on the number of people alone, but on the condition of the crops also." This dependence on a seasonal cycle is again characteristic of a premodern economy.[10]

The supply of salt. The essential premodern element affecting supply was fuel. The raw material of salt was brine — obtained from the sea, from wells or lakes, or by dissolving rock salt — from which mineral salt was produced by evaporation. Evaporation could be either natural, that is, by solar energy, or artificial, that is, by natural gas, coal, or various kinds of firewood, especially reeds.

Either form of evaporation had advantages and disadvantages. Solar power was free, but it was irregular and dependent on the weather, a disadvantage during the summer monsoon season in south China. Equally important was the problem of pollution: brine in the course of manufacture into salt had to spend from ten to fourteen days in the open air in shallow earthen beds, a process which afforded abundant opportunities for pollution by sand or other debris. Solar-evaporated or *shai* salt might be cheap, but as the customs report stated: "salt made in this way is always inferior to that made by boiling." [11]

On the other hand, the chief disadvantage of artificially evaporated salt was cost, both on account of the fuel itself, whatever this might be, and because of the large iron cauldrons used in the boiling process, which in Szechwan cost $30 and had to be replaced after seventy heatings. The production cost of boiled salt was at least double that of *shai* salt: what made it competitive was its quality and independence from climatic conditions.

It was essential, though, to keep the cost down by using local fuel supplies and to use a highly saline brine to keep the hours of boiling and hence the amount of fuel down to a minimum. At Tzu-liu-ching in Szechwan both problems were solved by the fortunate coincidence of exceptionally saline brine wells with natural gas wells. Another answer had been found in Huai-nan. Here the fuel was a special grass or reeds abundant in the vicinity. The raw material was "the water of the salt marshes along the seacoast, which is more saline than sea-water," [12] but its salinity was then intensified by treating it with the ashes of the reeds, which because they grew in the salt marshes had absorbed large quantities of salt. Thus the reeds, serving both as a fuel and as a salinity-intensifier, were a key element in the Huai-nan industry.

Availability of fuel was, in fact, the crucial factor in deciding which method of salt manufacture to adopt. Since no other locality enjoyed the advantages of Tzu-liu-ching, and since the cost of moving coal even a short distance was prohibitive, most salt in China was produced either by solar evaporation or by the Huai-nan method. This decision to use the cheapest form of energy was rational enough, but it produced countermodernizing side effects.

The choice of fuel isolated the salt industry, driving it into backward regions where land was available for evaporation beds or the growth of reeds. Except at Tzu-liu-ching, salt production in China was a rural industry not closely associated with any other industry. Indeed the salt trade positively repelled other industries: in Huai-nan, export of reeds, probably for soda manufacture, was forbidden to preserve the fuel supply, and Chang Chien found his plan for extending cotton cultivation opposed by the Liang-huai salt commissioner for fear it would interfere with the growth of reeds. Contrast this with seventeenth-century England where the use of coal brought the salt, glass, and alum industries together at Newcastle. The isolation of the salt industry in China because of its choice of fuel made its subsequent mutation into the servant of industrial chemistry all the harder.

The result of using primitive sources of power was to make the salt industry one with a low level of capitalization. Again, Szechwan with its brine wells costing up to 40,000 taels to sink, its firewells, pipelines, and teams of bullocks was an exception, for in general the salt industry in China required little fixed capital beyond the acquisition of land and, in the boiling regions, iron cauldrons. The Chinese salt merchant is often blamed for not re-investing his profits in his business, but the technical conditions of his industry gave him little encouragement to do so. Premodern in its consumer direction, the salt industry was no less so in its production methods.

Commerce: Limited Opportunities for Expansion

In 1900, the salt trade was still, as it had been in the eighteenth and nineteenth centuries, generally flourishing and institutionally sophisticated; in Saeki Tomi's description, it was one of the *zaibatsu* of imperial China. Behind this appearance of solidity, however, lay the stagnation characteristic of the premodern gabelle: what was premodern here was the limited capacity for expansion within the existing commercial structure. Most of the potential had already been realized, and three elements in particular had become obstacles to further development: the pattern of the routes followed, the role of the salt trade in China's economy, and the rate of turnover.

The pattern of routes. The formal pattern of the salt trade consisted of the twelve regional *yin-ti.* Chang Chien defined the *yin-ti* as "the Government prescribed monopoly selling districts, which must consume the salt from certain designated producing territories," and the term was variously translated as salt circuit, salt division, administrative division, territorial salt district, and salt administration. Here, H. E. Wolf's translation, salt division, has been adopted to avoid confusion with the ordinary circuit (tao) and district (hsien) of Chinese administrative geography.[13]

The salt divisions, which included the producing areas, by no means coincided with provincial boundaries; according to the customs commissioner, a salt division "does not always include the whole of the province in which salt is produced, and on the other hand, almost invariably overlaps provinces or parts of provinces other than the one supplying the product." Ho-tung division for example, whose source of supply was a salt lake in southwest Shansi, supplied southern Shansi, eastern Shensi, and western Honan. Northern Shansi, on the other hand, was supplied from salt lakes in northwest Shensi, Alashan, and the Ordos.[14]

Salt from one division was not permitted to be sold in another, and checkpoints were established on the main routes to prevent this. The real defense of the monopoly, however, was comparative cost and quality: the salt division was generally a natural monopoly area before being made a legal one. If the balance of costs was upset, for instance by a sharp increase in taxation in one division, then the checkpoints were often insufficient to prevent an influx of contraband salt, and the local authorities might petition for a redrawing of the divisional boundaries to legitimize the de facto situation.

For example, in Hunan prior to the Taiping rebellion, Kwangtung salt was supposed to be sold only in two chou and one hsien, the rest of the province forming part of Huai-nan salt division. The disruption of the Yangtze trade routes during the rebellion gave Kwangtung salt an opening in Hunan from which it could not be dislodged afterwards, and the Huai-nan authorities had formally to permit Kwangtung salt to be sold in competition with Huai salt in one prefecture, Yung-chou, and in the greater part of another, Heng-chou. Kwangtung salt, however, because of lower production costs and less taxation, remained cheaper than Huai salt in Hunan — 30–40 cash a catty compared to 52–56 — so its advance continued. By 1910 the whole of Heng-chou and another prefecture, Pao-ch'ing, had fallen to Kwangtung salt, and the Liang-kuang governor-general Chang Ming-ch'i memorialized for a redrawing of the boundaries to include Yung-chou, Heng-chou, and Pao-ch'ing in the Kwangtung salt division. Already in 1901, the customs commissioner believed that Hunan consumed as much Kwangtung salt as Huai salt.[15]

The twelve salt divisions in 1900 and the approximate amounts of salt they produced and distributed are listed in Table 1. A more succinct and analytical appraisal of this system, however, can be

Table 1. The salt divisions in 1900.

Division	Amount of salt distributed (piculs)
Fengtien	3,600,000
Ch'ang-lu	4,000,000
Shantung	3,000,000
Ho-tung	1,500,000
Huai-pei	1,400,000
Huai-nan	5,000,000
Liang-che	3,500,000
Fukien	1,300,000
Kwangtung	4,000,000
Szechwan	5,500,000
Yunnan	500,000
Northwest	500,000
Total	33,800,000

Source: Customs Report; District Reports, 1913–1917; Dane Report; Hosie, "The Salt Production and Salt Revenue of China."

gained by taking as the determining factor the availability of inland water transport. In the 1870's Ferdinand von Richthofen found "that freight by land is from twenty to forty times more expensive than it is by water," and in Dane's time, too, "salt transported by land by human carriers, by pack animals and by carts, even if it has paid no duty to the Government, cannot . . . compete with duty-paid salt transported by water." *Viae salariae* did exist in China, especially in the West,[16] but salt was preferably moved by water.

There was, first, a major axis along the Yangtze on which were centered the two leading salt divisions, Huai-nan and Szechwan, whose salines each produced between 5 and 6 million piculs of salt, in all, a third of China's total production. By this route, the Huai-nan and Szechwan salines supplied over two-thirds of the 8 million piculs of salt required by Anhwei, Kiangsi, Hupei, Hunan, and Kweichow, which produced none of their own.

There was, second, a minor axis: a complex of inland waterways linking Peking to Canton, via the Grand Canal and the Ch'ien-t'ang, Kan, and Pei rivers, and, to the north of Peking, prolonged by the line of the Liao and Sungari, the inland route for Manchurian salt. At either end of this route, were the third and fourth salt divisions, Ch'ang-lu and Kwangtung, each producing 4 million piculs of salt and together supplying nearly 3 million piculs to the saltless provinces of Honan, Kiangsi, and Hunan.

Finally, in the four quadrants between the two axes were the lesser salt routes, such as the Hsiao-ch'ing River in Shantung, the Min and Han rivers in Fukien, and the river and caravan routes of Yunnan and the northwest. Most of these were of local importance only, but two were of more than this: by the Huai River, Huai-pei exported nearly 1.25 million piculs to northern Anhwei and southern Honan, and by the West River a million piculs of Yüeh salt reached Kwangsi from Canton and Pakhoi.

This network distributed a large quantity of salt, but by 1900 its possibilities were fully exploited and it was beginning to stand in the way of future development. Ch'ang-lu salt, for example, was held back by the inadequacies of the Grand Canal as a link between north and south China and by the absence of a good water route to the northwest. Gustav Detring, the Tientsin customs commissioner, noted in 1901: "The Shansi and Mongolian salt is said to have a very bitter taste, and is therefore not so much valued as seasoned Chihli salt; and there is no doubt that if the latter could be transported by rail into those distant provinces, it would be preferred by the consumers." [17]

Similarly, the salt lakes of the far northwest, Dabasu and Kokonor, to which contemporary China claims to look for future supplies,[18] were unexploited in common with most of the northwest. Lack of transportation was the obstacle to development, and Richthofen's remarks were still in point: "As to the provinces of Shensi and Kansu, the benefits which would accrue to them, from an improvement of the means of intercommunication, would perhaps be even greater than those that would be enjoyed by Shansi and Honan." [19] The salt trade, like the rest of China, was thus due for reorientation, once the old dominance of the Yangtze could be broken by the development of modern communications.

The pattern of exchange. In medieval and renaissance Europe, salt, along with other bulk goods such as timber, wine, building stone, and cereals, was one of the principal trading commodities. Dutch commercial supremacy in the Baltic in the sixteenth and seventeenth centuries, for example, rested to a considerable extent upon control of the importation of salt from Setubal in Portugal and the bay of Biscay, and Setubal salt also played a substantial part in the sixteenth-century trading empires of Lisbon and Seville. With the development of a modern economy in Europe, however, while the salt trade expanded in volume, its relative importance declined, as other commodities came to the fore.[20]

In China in 1900, on the other hand, the salt trade retained its premodern eminence, both quantitatively as a percentage of total trade and as an economic activator. Thirty million piculs of salt were produced and sold in China and, at an average price of 40 cash a catty and at 1,200 cash to the tael, were worth 100 million taels. This was over half the value of China's total foreign exports passing through the Maritime Customs in 1901 (169 million taels). At this time the value of native opium traded in China was 130 million taels, and that of the interregional trade in rice, 100 million taels. Salt also remained a key commodity in the pattern of exchange. In Manchuria, the boats bringing down beans, grain, and tobacco loaded salt for their return journey to the interior; a harvest surplus in Honan was carried to Shansi, a chronic grain-deficit area, by the returning salt carts from the An-i-hsien salt lake; in Chekiang, in the Ch'ien-t'ang valley, salt was exchanged for tea, hams, and building materials; in Kwangsi, the salt junks from Canton, "the best kept and best preserved boats on the river," carried firewood for their return freight; while in the southwest, "Two-thirds of the total quantity of cotton imported into Yunnan from the Burmese Shan states and French territory bordering on the Mekong is paid for in salt . . . Salt is also the principal medium of exchange in the tea trade." [21]

As in early modern Europe, these exchanges fell into one of two patterns. On the one hand, salt could act as the instrument of urban domination over some less well-developed region, "un des aspects les plus criants de la domination économique urbaine": [22] this was the case with the sixteenth-century Venetian salt trade in relation to the Adriatic hinterland, for example. In China, this was the role of Yangchow salt in relation to primary-producing Hunan and Anhwei and of Canton salt in relation to rice-exporting Kwangsi.

On the other hand, salt could act as the instrument of rural compensation, the one product of an underdeveloped area needed by a more highly developed one: this was true of Brouage salt in relation to Paris and Bordeaux in eighteenth-century France, for example. In China, this was the position of the Huai-nan salines in relation to the Yangtze delta cities, or of Alashan Mongol salt in relation to the cities of the northwest, such as Sian, or of Pakhoi and Hainan salt in relation to Canton. Some salt-producing areas, indeed, occupied both positions. Tzu-liu-ching, urban in relation to primary-producing Kweichow, was rural in relation to Hankow, and Szechwan salt exports joined opium in meeting the cost of manufactured imports from down river.

Neither role offered much opportunity for constructive expansion. The modern pattern of urban-rural relations is one of mutual development rather than simple domination or compensation by primary products. An alimentary salt trade could play little part in this sort of development, since it could not set off developmental reactions; that is, its own market had a definite saturation point.

The turnover factor. The third factor hindering the expansion of trade was the rate of turnover. Given the premodern character of the communications network, turnover in the salt trade was bound to be slow, but it was further retarded by the institutional framework within which it had to operate.

Superficially, the salt trade was carried on under a bewildering variety of institutions: the salt divisions differed one from another in organization, and even within a division many different arrangements were to be found. In Szechwan, the predominant system was "official transport and merchant sales" (*kuan-yün shang-hsiao*), the officials taking the leading role in the trade, while in Huai-nan it was "official supervision and merchant sales" (*kuan-tu shang-hsiao*), privileged merchants conducting the trade under official supervision only. Both divisions, however, used other systems as well: in Szechwan, 55 percent of the salt exported to Hupei was brought down under merchant, not official, transportation, while in the Wan-an section of Huai-nan, in two districts of Anhwei close to the Kiangsu border, Ch'u-chou and Lai-an hsien, the "official transport and official sales" (*kuan-yün kuan-hsiao*), or complete government monopoly system, was in force. Beneath the diversity, however, there was a common substratum, the differences between the systems turning out to be less striking than their similarities. In Szechwan, on the Hupei run, the official transport office (*kuan-yün chü*) in fact operated by subcontracting the transportation of its salt to salt merchants (*yen-hao*), who were already in the business on their own account and who thus approximated in role the transport merchants (*yün-shang*) of Huai-nan; while in Huai-nan, "The *yün shang's* business transactions are under official control to such an extent that he is practically an official servant." A combination of close official control with subcontracting to privileged merchants characterized all the systems.[23]

The effects of what was in practice a general "official supervision and merchant operation system" (*kuan-tu shang-pan*) had not always been harmful. On the contrary, official control had brought capital into the business, both directly in the form of government funds under the "official transport" system (*kuan-yün*), and indirectly, by encouraging merchant investment through the grant of monopoly privileges, under the "merchant transport" system (*shang-yün*). This was of particular benefit in outlying areas, such as Kweichow and Kirin and Heilungkiang, where the trade might not have proved sufficiently profitable to stimulate investment.

Even Dane, a severe critic of the whole system, was prepared to admit that it had its beneficial side in these regions. Indeed, it is not even true to say that all areas of the salt trade were depressed in 1900. Yüeh salt from Kwangtung was expanding its sales in Hunan, thanks to low production costs and low taxation, while Ch'uan salt, thanks to its superior quality, was finding a market as far away as Honan, despite the fact that it cost twice as much as the local Ch'ang-lu salt.

However, in the premier salt division of the empire, Liang-huai, signs of malaise were already visible. All was not well with the Huai-nan trade at the beginning of the twentieth century. In 1901, H. E. Wolf, salt likin collector at Ta-t'ung and perhaps the best informed on salt affairs of all the customs commissioners, headed one of his sections of the customs report "stagnation in salt trade"; in the previous year, Liang-kiang governor-general Liu K'un-i had memorialized that "the Huai-nan merchants are in great distress"; and in 1903, his successor Chang Chih-tung memorialized, "Huai salt is very weak and reconstruction urgent." [24]

Western observers, looking at Huai-nan salt division from a modern standpoint, had different diagnoses for its difficulties. On the one hand, Wolf, in a long excursus in the customs report, developed the argument that the essential trouble was overproduction: "The difficulties under which the trade in legitimate salt, and in sympathy therewith the salt revenue, in the Liang Huai salt division labour are briefly stated as follows: the root of the evil is the continued over-production of salt by the *ch'ang* [lit., 'yards,' the producing centers] of the division." [25]

Overproduction, he argued, led to delays in marketing, since salt was sold and transported strictly in accordance with the order in which it was produced and lodged in the official depot. Serious delay took place both at Shih-erh-wei, where the producer (*ch'ang-shang,* lit., "yard merchant") sold to the transport merchant (*yün-shang*), and also at the up-river port where the transport merchant sold to the provincial distributor (*shui-fan,* lit., "water merchant"). Wolf wrote in 1901: "The salt now brought to Wan-an is that of the season, or *kang,* 1890. After landing it here the salt can only be sold again when its due turn comes round — not until after 12 to 14 months storage." [26]

As the turnover period lengthened, profit per unit of capital over time fell, even though markup per unit of salt sold remained the same, or even increased. Wolf reckoned: "The profits of the *ch'ang shang* would seem to be about 50 per cent. on the net value of his salt in depôt, but, on taking into consideration the length of time he has to wait until he realises the capital sunk in the salt, his profits dwindle to about 10 per cent. per annum." Similar considerations applied to the transport merchant. To escape this situation, salt merchants resorted to smuggling, but this only slowed down legitimate sales still further. The true remedy, Wolf believed, was the "removal of the stagnation at present prevailing in the salt trade by regulating the output of salt at each *ch'ang*." [27]

Dane, on the other hand, believed that the problem was one not of overproduction, but of underconsumption: "It is probably no exaggeration to say that in all places, where monopolies of transportation and sale exist, the demand for salt is not fully met." [28] It was this underconsumption which led to accumulation of stocks and stagnation of trade.

The reason for it was that "Monopolists can make the same amount of profit by selling a comparatively small quantity of salt at a high price as they can by selling a large quantity at a low price; and as it costs them less to adopt the former course (less capital being necessary for the purchase and transportation of salt), it is certain that they will adopt it." Prices, he admitted, were fixed by the government, but "the monopolist can always afford to corrupt Government officials by bribes and presents, or even the Government itself by contributions when funds are urgently needed, and to thereby induce them to increase the sale prices of salt." The official price, Dane believed, gave the merchants "enormous profits," and these, in combination with heavy taxation, reduced sales.[29]

It was indeed the case that in Huai-nan nearly 78 percent of the consumer price of salt was made up of taxation and merchant profit, a figure so similar to the 81 percent found by Fernand Braudel in cities supplied with salt by Venice in the late sixteenth century that it may perhaps be regarded as characteristic of premodern salt monopolies.[30] It was the monopolies therefore, according to Dane, which were inhibiting the trade, and the remedy was competition: "Monopolies and trade rings should, as far as possible, be prevented"; and again, "If salt is transported and sold in large quantities, the price will fall, consumption will increase and the revenue of the Government will increase proportionately. The Chinese system of monopolies is a complete negation of these principles." [31]

Both Wolf's and Dane's explanations of the stagnation of trade in Liang-huai contain undeniable truths: the market for alimentary salt, unlike the market for chemical salt, has a low ceiling, so that overproduction is a real danger, while monopoly, of its nature, must hold the possibility of restriction.

Both, however, leave something unexplained, for, on the one hand, it is doubtful that alimentary salt had genuinely reached a ceiling in China in 1900. Wolf in one passage admitted this, at least for Anhwei: "In the present unhealthy condition of the salt administration, and with excessive smuggling still on the increase, 845 *p'iao,* or licenses, for the Wan-an Huai salt district are regarded as being far in excess of actual requirements, but, as will be shown later on, 845 *p'iao* at 120 *yin* may well be used per year under an honest and efficient administration." [32] Wolf's hopes for an increase in consumption were in fact justified under Dane's reformed administration.

On the other hand, the monopolies were not inevitably restrictionist. As has been seen, in earlier days and perhaps still in Kweichow and Kirin–Heilungkiang, monopoly brought capital into the salt business which it would otherwise have been unable to raise. It might pay a monopolist to sell less rather than more, but it was the state of the market, the incidence of taxation, comparative costs, and so forth which would decide this, not the built-in nature of the monopoly itself, and it is ultimately these varying factors whose bearish tendencies in 1900 need explanation.

A further reason for the decline in investment in salt was one mentioned by Wolf, but regarded by him as secondary to overproduction. This was the factor of turnover time. Profit in the salt trade, as elsewhere, was relative to time: the longer it took to move salt from the *ch'ang* to the consumer, the greater must be the profits to make it worthwhile in relation to other forms of investment. R. de Luca, Hupei salt likin collector, emphasized this element in commenting on the transport merchant's profit, "the relative importance of which is, of course, entirely dependent on the degree of speed with which the salt can be disposed of, accounts settled, and a new venture undertaken." If turnover time increased, not only would annual profit fall, but outlay might increase through higher storage costs, bigger stocks, and so on, and a salt merchant might well decide to limit his investment in the trade. In Ch'ang-lu, Dane found that "The great majority of the merchants did not transport salt themselves and merely sublet their rights of transportation and lived, like absentee landlords, upon the rents which they received." Indeed, following the Boxer rebellion, they borrowed 7 million taels from European traders and tried to diversify their business activities, though in fact they only incurred new losses and debt.[33]

Wolf was right to stress the factor of delay. However, it might be argued that the reason for it was not overproduction, for this was relative to the merchants' purchases. Rather, it was the result of supervision by a premodern bureaucracy whose intervention in the trade at all times had harmful as well as beneficial effects, intensifying as the cost of government rose.

Every aspect of the salt trade required official sanction and the issue of permits. As an illustration, let us consider the marketing of a consignment of salt from the Huai-nan salines to an up-river port such as Hankow or Ta-t'ung. The yard merchant (*ch'ang-shang*) at the salines had to report his output of salt daily to the salt receiver (*ch'ang-kuan*), and could carry salt for sale to Shih-erh-wei only on the requisition of the Huai-nan general office (Huai-nan tsung-chü) at Yangchow, transmitted through the salt receiver; when he received this, he had to apply back to Yangchow for a transport permit (*ch'ung-yen chih-chao*) and to his local branch office (*fen-ssu*) for a cargo certificate (*ts'ang-tan*).

Meanwhile, his prospective customer, the transport merchant, had to petition his provincial sales office (*tu-hsiao tsung-chü*) for an official communication (*tzu-wen*) to the authorities in Yangchow. Armed with this, he arrived at Yangchow, paid half the price and tax on his salt, went to Shih-erh-wei and made his bargain with the yard merchant, and then returned to Yangchow to pay the balance. The Huai-nan general office then issued a delivery permit (*fa-yen chao*) allowing him to take and load his salt; but before sailing, he had to obtain a cargo certificate from the storehouse officials at Shih-erh-wei, an official communication from Yangchow for the provincial sales office, and, most important, a passport (*hu-p'iao*) in triplicate, with the butt remaining at Yangchow, the middle portion traveling with the consignment, and the left-hand portion (*tso-chao*) sent ahead to the up-river port. Similar procedure awaited him there.

The stagnation of trade in Huai-nan was indeed a complex matter. In addition to red tape and to Wolf's and Dane's factors of overproduction and underconsumption, there was also the problem of salt smuggled in from the surrounding salt divisions: Hotung, Huai-pei, Liang-che, Kwangtung, and Szechwan. The existence of an illicit trade in salt, parallel to the legal traffic, was a well-established feature of the old regime, and one which took a variety of forms: tax evasion by the transport merchants using the regular routes was the most prevalent, followed by the transport of salt along uncontrolled routes both by recognized traders and *banditti,* along the mountainous Pakhoi–West River route, for example, and by the export of duty-paid salt from a low-tax salt division into a highly taxed one, such as Huai-nan.

In the short run perhaps, a decline in legitimate traffic might be owing largely to an expansion of the illicit trade, though B. D. Bruce of the Anhwei salt likin collectorate seemed to reject this as an explanation of Huai-nan's difficulties when he commented that "Smuggling is reported to be not so prevalent as in the immediate past." [34] In the end, however, the illicit trade was subject to the same inhibiting factors as the legal: inadequate communications, inelastic patterns of exchange, and a slow rate of turnover. The trade in smuggled salt was not in the long run more dynamic than the legal trade, and, rather than growing at the latter's expense, tended far more to parallel its market fluctuations. In the last analysis, the major responsibility for the stagnation of trade in Huai-nan, it may be suggested, lay not with the competition from smuggled salt, but rather with the bureaucratic framework which, more than anything else, initiated the cycle of slower turnover, accumulation of stocks, decline of profit, and eventual disinvestment.

The Fiscal Aspect: Assessment Not Comprehensive, Receipts Not Consolidated

Despite a genuine concern with the social aspects of salt, the primary function of the salt administration was fiscal. Here, its salient unmodern feature was a lack of systematization; indeed, it was characterized by a proliferation of anomalies which made one customs commissioner despair of understanding it: "The system of taxation and distribution is very complex, and appears to the uninitiated, like myself, as if it were made purposely to defy a thorough grasping of its details."[35] In a premodern context, this complexity seems to have been an inevitable element in the process whereby the bureaucracy accommodated itself to local conditions which were beyond its control; but by modern standards, it was simply inefficiency, particularly in the two key activities, assessment of tax and its allocation after collection.

Assessment of tax. Here, the chief inadequacy was the failure to tax all salt. A large part of the salt trade went unrecognized by the authorities and hence untaxed, a clandestine penumbra ranging from legalized tax evasion to cloak-and-dagger "owling" (*hsiao*). Dane wrote: "The inequality of the taxation was in fact one of the greatest blots on the administration."[36]

This inequality took various forms. Like most traditional governments, the Chinese bureaucracy had to tax in accordance with inability to avoid rather than ability to pay. Consequently, tax was levied not on salt as such, but on salt traveling long distance by water, since that alone was bound to follow well-defined routes and to pass through centers where avoidance would be difficult. Dane explained the system: "The founders of the Chinese policy grasped the supreme importance of the transportation by water. In the areas which had to be supplied with salt by water, because the cost of transportation by land would have been absolutely prohibitive, the administration was highly organised." The result, however, was that "The people to whom salt was naturally dear owing to the distance at which they lived from the salt sources, had to pay the highest taxes. The people, who could best afford to pay a salt tax, namely, those to whom salt was naturally cheap, escaped almost entirely scot free." Inasmuch as the lightly taxed regions included some of the richest areas in China, the system was both unfair and inefficient in terms of salt tax at least.[37]

Szechwan and Huai-nan illustrate this. Szechwan salt division consisted of four parts: (1) *p'iao-an,* Szechwan districts supplied with *p'iao* or land-borne salt; (2) *chi-an,* Szechwan districts consuming *yin* or waterborne salt; (3) *chi-ch'u,* half the province of

Hupei and a small portion of Hunan supplied from Tzu-liu-ching via the Yangtze; and (4) *pien-an,* Kweichow and northern Yunnan, also so supplied. Table 2 shows the salt consumed in each of these

Table 2. Salt releases and revenue in Szechwan before reform.

Area	Salt releases (piculs)	Percentage of total	Revenue (taels)	Percentage of total
P'iao-an	1,375,000	25	871,961	12
Chi-an	1,793,000	33	899,999	14
Chi-ch'u	990,000	18	2,375,106	34
Pien-an	1,342,000	24	2,740,917	40
Szechwan	5,500,000	100	6,887,983	100

Source: Dane Report, p. 144; District Reports, 1913–1917, p. 68; Customs Report, p. 42. Besides Kweichow and parts of Yunnan, the *pien-an* also included Szechwan districts adjoining these provinces. For the quotas of salt supplied to these "protective districts," see District Reports, 1918, pp. 79, 82, and District Reports, 1913–1917, p. 94. The *chi-ch'u* tax figure includes the levy made at Ichang. All tax figures are averages based on the period 1903–1910.

parts and the taxes received from it: nearly three-quarters of the tax came from 42 percent of the salt, the export trade to Kweichow and Hupei. Szechwan itself, in Richthofen's view the most prosperous province in China, went comparatively untaxed.

In Huai-nan, there was a similar contrast between the lightly taxed Yangtze delta, the northern part of which lay within Huai-nan, and the highly taxed up-river provinces. Huai-nan was divided into two parts; the four Yangtze *an* (lit., "ports," here meaning "consumption areas") in Anhwei, Kiangsi, Hupei, and Hunan on the one hand, and the *shih-an,* the local consumption area in Kiangsu nearer to the source of supply. While the four *an* each had a separate sales office, they were not further subdivided for tax purposes, but the *shih-an,* on the other hand, was further subdivided for fiscal purposes into (1) the region near the salines, (2) Yangchow itself, and (3) Nanking.

Salt for any of the four *an* paid tax twice: a smaller installment at Yangchow when the salt passed from the yard merchant to the transport merchant, and a larger amount at the up-river port on passing from the transport merchant to the provincial distributor. The rates of tax for all four *an* were approximately the same, those for O-an, or Hupei districts supplied with Huai salt, being at

Yangchow 1.3066 taels a *yin* (1 *yin* = 7.50 piculs) and at Hankow 10.1235 taels a *yin,* making a total tax of 11.4301 taels a *yin.*

In the *shih-an,* on the other hand, salt for the salines' region and for Yangchow paid tax only once, at Yangchow, at the rates of 1.3703 and 2.0280 taels per *yin* respectively, and, while salt for Nanking paid an additional likin on arriving there, the total tax came only to 4.6760 taels a *yin.* This difference in tax was reflected in the consumer price for salt: in the four *an,* it was 50 cash a catty, in the *shih-an,* 22.

The Yangtze delta was undertaxed in a second way, also widely prevalent in China. The official quota of salt for the *shih-an,* on which tax was paid, was only 381,360 piculs, not enough for a region containing two major cities, Yangchow and Nanking, as well as several lesser ones along the Grand Canal. Smuggling made up the difference, so that only a fraction of the total salt consumed was taxed at all, unlike the four *an* where the quota was more realistic.

South of the river, where the rest of the delta belonged to Liang-che salt division, the position was even worse. Tax rates, though higher than in the *shih-an,* were below those in the four *an* and the quota, "namely, 408,720 piculs, was wholly insufficient for the requirements of the population." The Yangtze delta region, the most thickly populated area in China, probably consumed over 2,000,000 piculs of salt, and only a third of this paid tax, and then only at modest rates.[38] Other urban areas close to salines, such as Peking, Tientsin, and Canton, were treated in the same way. The salt tax was thus primarily a tax on salt carried long distance by water, and areas close to the salines, even if rich and populous, paid little.

Yet another inequality in the method of assessment was that not even all salt transported long distance by water paid tax. Tax was levied according to the *yin,* or the consignment of bags of salt, though the weight of the *yin* and the number of bags varied from division to division: in Huai-nan, for example, the number of bags was eight and the weight of the *yin* totaled 7.50 piculs or 750 catties, while in Szechwan, the *yin* consisted of fifty bags and weighed either 80 or 100 piculs depending on whether the salt transported was cake or granular. In practice, however, as the Hankow customs commissioner pointed out, "the *yin* is invariably of greater importance as a salt measure than the weight in catties, it being almost impossible to keep exact tally of actual weights owing to the nature of salt itself, subject as it is to variation from climatic conditions."[39] A certain imprecision was thereby imparted to the assessment.

As salt was liable to waste en route through humidity, merchants were allowed to transport a certain amount of tax-free salt in addition to the salt on which they had paid duty, to compensate for possible losses. Though fair in principle, in practice this wastage allowance (*lu-hao*) became the principal vehicle for smuggling. The existence of legitimate excess weight in the form of the wastage allowance afforded the merchant further latitude to carry more salt than his entitlement. The connivance of officials could be purchased and smuggling generally took this form, the illicit trade flowing in the same channels and being conducted by the same people as the licit, the serious smugglers being not *banditti* but the transport merchants themselves. As Dane saw: "This smuggling by the recognised transportation routes is the smuggling which causes the most serious loss of revenue . . . the monopolists were allowed to entertain and pay private Salt Police, who assisted to some extent in the prevention of smuggling by outsiders, but who, so far as the Government was concerned, were merely agents for the promotion of smuggling by the recognised merchants."[40]

Receipt of tax. The other premodern element in the fiscal side of the salt administration was the failure to pay all taxes received into a single consolidated account at the disposal of the central government.

Finance in imperial China was decentralized, and the salt administration with it. There was no single salt tax: in Ch'ang-lu, for example, before the foreign inspectorate's reforms there were twenty-four heads of assessment; at Ta-t'ung in 1901, there were ten; at Ichang, seven; at Hankow, ten; and in parts of Shantung as many as forty-three. Some were designated for particular purposes from the outset; in Ch'ang-lu, there were special salt taxes for the expenses of the imperial board of astronomy, for Jung River works, and for road maintenance, while at Kiukiang there was a levy of 0.010 taels per *yin* for the local foundling hospital. Others were simply assigned to various treasuries for their general purposes. Thus, in Hunan in 1901, of the 1,497,000 taels collected in likin on 1,152,000 piculs of Huai salt imported into the province, 633,600 were assigned to the Hunan authorities and 864,000 to the Liang-huai salt controller, who was a subordinate of the Liang-kiang governor-general at Nanking.[41]

In all areas, only a fraction of the gross collection was reported to Peking, and of this "reported salt revenue" only a part was genuinely at its disposal. The reported collection, 13,490,000 taels, was a matter of public knowledge, reported in the press; the gross collection is harder to calculate since the customs report does not give complete tax lists for all salt divisions, especially in the south, but 24 million taels would be a conservative estimate for 1900.[42]

The amount genuinely at Peking's disposal is even more difficult to assess, inasmuch as the apportionment between Peking and the provinces was not a fixed one and varied according to political circumstances. The customs commissioner at Yochow suggested that "some *Tls.* 9,000,000 are accounted for to the Board of Revenue,"[43] and this is some indication of what was available to Peking. Table 3 lists the three collections — gross, reported, and available — for the key years between 1900 and 1922.

Table 3. Salt revenue, 1900–1922: gross collection, reported collection, collection available to Peking (millions of taels).

Year	Gross collection	Reported collection	Available collection
1900	24	13	9
1908	29	17	9
1911	40	20	9
1914	40	40	40
1918	47	47	37
1919	53	53	36
1922	57	57	36

Source: CSL:HT, 61:3–4; Customs Report, pp. 32, 69; *YWKMS*, p. 5; "La Gabelle du sel," p. 436; Central Salt Administration Accounts. For gross collections in 1900, see below, note 42.

Two examples will illustrate how the system worked. Table 4 sets out Hankow taxes on Huai salt imported into Hupei where there were ten heads of taxation totaling 10.1235 taels a *yin*. Nine treasuries shared the 1,343,464 taels collected on Huai salt at

Hankow: three belonging to the Hu-kuang governor-general (Wuchang salt taotai, Hupei ordnance department, Hupei provincial government), three belonging to the Liang-kiang governor-general (Liang-huai salt controller, Nanking defense treasury, Hupei sales

Table 4. Hankow revenue allocations, 1900.

Tax	Rate per *yin* (taels)	Office allocated to
Hupei likin (*Ch'u-li yin*)	1.8000	Wuchang salt taotai
Additional salt tax (*chia-k'o yin*)	1.2130	Liang-huai salt controller
Junk duty (*ch'uan-shui yin*)	0.1400	Kiukiang native customs
Huai likin (*Huai-li yin*)	2.2781	Nanking defense treasury (Chin-ling fang-ying chih-ying chü)
"Grant-in-aid" tax (*hsieh-hsiang yin*)	1.0489	Huai army transport department (Huai-chün chuan-yün chü)
River defense tax (*chiang-fang ching-fei*)	0.7250	Hupei ordnance department (Hu-pei ch'iang-p'ao chü)
Coast defense tax (*hai-fang ching-fei*)	0.7250	Nanking defense treasury
New army tax (*lien-ping hsin-hsiang*)	1.0625	Hupei provincial government, 75 percent; Ching-chou tartar-general, 25 percent
Office charges (*chü-yung*)	0.3750	Hupei sales office (*tu-hsiao tsung-chü*)
Miscellaneous exactions	0.7560	Hupei sales office

Source: Customs Report, pp. 107–109, 111–112.

office), and three independent of the provincial authorities (Ching-chou tartar-general, Kiukiang native customs, Huai army transport department). Furthermore, out of the total collection, 500,000 taels had to be handed over by the other treasuries to the Hupei salt likin collectorate as the Maritime Customs quota for servicing the Anglo-German loan of 1898. De Luca, the commissioner in charge, does not specify which items were earmarked for this:

probably they were the Hupei and Huai likins which would have brought in rather more than the required amount.

The second example, tax collections at Ta-t'ung on Huai salt imported into central Anhwei (Table 5), is more typical, since

Table 5. Ta-t'ung revenue allocations, 1900.

Tax	Rate per *yin* (taels)	Amount (taels)	Office allocated to
Good-will tax (*pao-hsiao yin*)	1.8715	139,688	Chiang-ning treasurer and Nanking defense treasury
Antismuggling tax (*ch'i-ssu yin*)	0.1970	14,706	Anhwei sales office (*tu-hsiao tsung-chü*)
Office expenses (*chü-fei yin*)	0.1539	11,487	Anhwei sales office
Merchants' tax (*t'ieh-shang-p'ing yü-yin*)	0.0154	1,148	Anhwei sales office
Steamboat fuel tax (*lun-ch'uan t'an-fei yin*)	0.0390	2,940	Anhwei sales office
Price increase No. 1 (*chia-chia yin*)	1.000	82,800	Nanking treasury (*chih-ying chü*)
Price increase No. 2	1.000	82,800	Governor of Anhwei
Salt likin (*yen-li*)	3.8988	322,820	Salt likin collectorate
Meltage fee (*huo-hao yin*)	0.0327	2,440	Anhwei sales office
Bankers' discount	0.0093	771	Official bankers

Source: Customs Report, pp. 136–140.

only one governor-general was involved. There were ten heads of taxation amounting to 8.2176 taels a *yin* bringing in a total of 661,600 taels which seven treasuries shared among them. All, however (except the salt likin collectorate which, with the collectorates at Hankow and Ichang, acted for the Maritime Customs under the 1898 loan agreement), were under the control of the Liang-kiang governor-general.

Even his control, however, was not complete. The salt administration, like much in imperial China, worked through an informal system of checks and balances. On the one hand, each of the Liang-kiang treasuries involved represented competing institutional vested interests, with which the governor-general could not arbitrarily tamper once their salt appropriations were fixed. On the other hand, though he enjoyed a wide measure of autonomy in salt affairs as elsewhere, the governor-general was still responsible to the court. Any plans he might make for increased taxation, or for new uses for old revenues, had to be cleared with Peking. In 1899 for example, Chang Chih-tung memorialized that the cost of moving 30,000 *shih* of Hunan tribute rice to the capital be assigned to the Ichang salt likin receipts, and in 1901 Yun-kuei governor-general Ting Chen-to memorialized to increase the salt price to pay for a local militia force.

For its part too, the court could make new demands on the salt revenues: a decree of 1899, arranging to meet a 2.5 million tael bill for Manchurian border defense forces, demanded quotas of 180,000 taels and 144,000 taels from the Szechwan and Huai salt likin offices respectively. The line between what Peking controlled, and what the provinces controlled, was not a fixed one, but fluctuated according to the situation and bargaining strength of both sides. New imperial demands were not always met. In 1901, when the salt divisions were being asked to contribute to the Boxer indemnity, Liang-kiang governor-general Liu K'un-i refused the board of revenue's request to increase the price of Huai salt in Hupei and Hunan by 4 cash a catty, pointing out that the price had already been increased substantially to meet earlier indemnities and suggesting instead an increase of only 2 cash a catty to spare the merchants further hardship. Similarly, in 1903 Yüan Shih-k'ai argued that the Ch'ang-lu merchants were already paying a price increase totaling 700,000 taels and advised against any further increase.

This absence of clear-cut rights over revenue was another facet of the lack of systematization prevailing throughout the fiscal level of the salt administration. The traditional gabelle collected a large revenue, the life blood of several imperial institutions, among them the Huai army, Li Hung-chang's political base, and the imperial household department (Nei-wu fu), which was the empress-dowager's. Yet, by modern criteria, its standards of efficiency were too low: too much salt went untaxed, too much revenue was dissipated into too many treasuries, and control over them was too much vested in the provincial governors-general. Decentralization of itself is, of course, not unmodern — a modern society may decide that certain functions are best conducted by small units — but the unorganized, involuntary decentralization of the traditional gabelle was incompatible with the requirements of modern fiscal administration.

Administration: An Unsalaried, Generalist, and Politically Compromised Service

The salt administration was distinct from the rest of the bureaucracy not in organization, personnel, or methods, but only in function. It shared with all of the bureaucracy a common premodern character, a lack of "rationalization" in Max Weber's sense, in that its officials were corrupt and insufficiently specialized; and, to compound these inadequacies, they were too much involved in politics.

Corruption. Corruption is a misnomer, a pejorative description for a system with its own code of behavior. It is convenient, however, as it points to an essential difference between modern and traditional bureaucracies. The latter, both in Europe and Asia, met their running costs not by fixed salaries and audited expense accounts, but by fees, unofficial exactions, commissions, and deductions from the revenue. Such a system is difficult to regulate and is apt to absorb an excessive proportion of the gross revenue: in France in 1683, even after Colbert's reforms, expenses still absorbed 22 million livres out of a total gross revenue of 119 million livres. Table 6 gives the percentage of revenue in China used up in collection costs in each salt division in the first year of the foreign inspectorate, when much of the ancien régime still survived.

Individually, salt officials could make large fortunes, the official salary being only part of the total emoluments. In Kwangtung, the salt controller (*yen-yün shih*) at Canton was estimated to make 300,000 taels a year, while his second-in-command, the assistant salt controller (*yün-t'ung*) at Ch'ao-chou, in charge of the export trade to Kiangsi and western Fukien, made 100,000 taels.[44]

Table 6. Collection costs in the first year of the foreign inspectorate.

Division	Year of establishment of inspectorate	Cost as a percentage of revenue collected
Ch'ang-lu	1913	9.34
Fengtien	1913	33.68
Shantung	1913	7.43
Ho-tung	1913	26.82
Liang-huai	1913	47.36
Liang-che	1913	14.59
Kwangtung	1913	7.30
Szechwan	1914	10.71
Yunnan	1914	21.83

Source: District Reports, 1913–1917

In Yunnan, a poorer salt division, the post of salt taotai (*yen-fa tao*) was "said to be worth about 40,000 taels a year," while his three leading subordinates, the salt inspectors (*t'i-chü*) in charge of the principal salt-producing districts of Hei-yen-ching, Pai-yen-ching, and Shih-kao-ching, had incomes ranging between 14,000 and 30,000 taels a year, though their official salaries were only 800 taels a year.[45]

The high profits from officeholding in the gabelle had two undesirable consequences: multiplication of posts and rapid changeover of officials. A memorial of the Grand Secretariat, just before the revolution of 1911, pointed out that the ratio of officials to revenue collected was much higher in the gabelle than in the land tax and that many salt offices were redundant: "every year one official relieving another, the good just following precedent in the conduct of business, the bad embezzling and getting into deficit." In Anhwei, at the time of the customs report, the general manager (*tsung-pan*) of the provincial sales office at Ta-t'ung held office for only one year, and when he left office, so did all his subordinates both at Ta-t'ung and at the *fen-chü,* since they were his appointees. Again, in Ch'ang-lu, seven occupants of the post of salt controller are mentioned in the *shih-lu* between 1900 and 1911. It is difficult to resist the impression that many posts in the salt administration in 1900 were exploitive rather than functional in character, and that something like a "spoils system" existed.[46]

Lack of specialization. It may be that the role of the specialist in modern bureaucracy has been exaggerated in the tradition of thought descending from Weber. In many modern states, England for example, and in most communist countries, top civil servants retain a generalist character. Some degree of specialization, however, especially in fiscal institutions such as the salt administration, is essential in modern government. This is particularly the case if, as in imperial China, there are no ancillary, specialized professions such as accounting and law whose members can be called in to give technical assistance to the generalists.

In 1900, the majority of ranking members of the salt administration were entirely unspecialized, their membership in the gabelle usually being a brief episode in their wider careers as generalists in the imperial service as a whole. This conclusion is based on Table 7, which shows the promotions of salt officials recorded in the *shih-lu* between 1900 and 1911.[47] From the table it follows that either the heads of the salt divisions were not specialists, or that provincial judges were not; both posts formed part of the normal *cursus honorum* of a successful official in imperial China, such as Yang Wen-ting, who went on to become governor of Hunan, and En-ming, who became governor of Anhwei. The promotion from salt commissioner (*yen-fa tao,* 4A, or *yen-yün shih,* 3B) to provincial judge (*an-ch'a shih,* 3A) was, perhaps, a critical one in a man's career, distinguishing, in British parlance, the "flyers" from the less talented officials, since the *shih-lu* have few records of appointments to salt commissionerships, presumably because they were of insufficient historical importance.

Only in a minority of cases do the careers of salt officials in early twentieth-century China suggest the salt expert or, more generally, the fiscal or economic expert. Chou Hsüeh-hsi for example, former Ch'ang-lu salt controller, was given first-grade official rank for his services to technology and served as general manager to Yüan Shih-k'ai's Lanchow coal mining company before he became minister of finance in 1912. Yang Tsung-lien, another Ch'ang-lu controller, had been a customs taotai at Hankow and had managed a number of modern-style enterprises for Li Hung-chang before taking control of the spinning works. K'o Feng-shih reappears later in the *shih-lu* as superintendent of the opium excise, and Yen An-lan, described by Dane as "a Salt Commissioner of much ability and experience," had been a secretary at the ministry of finance

Table 7. Promotion of salt officials, 1900–1911.

Original appointment	Name	Subsequent appointment
Liang-huai salt controller	K'o Feng-shih	Kiangsi provincial judge
Liang-che salt controller	Shih Chieh	Chekiang provincial judge
Fukien salt taotai	Yang Wen-ting	Fukien provincial judge
Ch'ang-lu salt controller	Yang Tsung-lien	Manager, Shun-chih machine spinning office
Liang-huai salt controller	Ch'eng I-lo	Kwangtung provincial judge
Liang-huai salt controller	En-ming	Kiangsu provincial judge
Shantung salt controller	Ying Jui	Hunan provincial judge
Liang-che salt controller	Hsin Ch'in	Chekiang provincial treasurer
Ch'ang-lu salt controller	Lu Chia-ku	Chihli provincial judge
Kwangsi salt taotai	Wang Jen-wen	Kwangtung provincial judge
Liang-che salt controller	Ts'ui Yung-an	Chekiang provincial judge
Kwangtung salt controller	En-lin	Imperial mausolea brigade general
Ch'ang-lu salt controller	Ling Fu-p'eng	Metropolitan prefect
Fukien salt taotai	Lu Hsüeh-liang	Fukien provincial judge
Liang-che salt controller	Hui Shen	Chekiang provincial judge
Liang-huai salt controller	Chao Pin-yen	Kwangtung provincial judge
Liang-huai salt controller	T'ang Shou-ch'ien	Yunnan provincial judge
Wuchang salt taotai	Ma Chi-chang	Hupei provincial judge
Kwangtung salt controller	Ting Nai-yang	Metropolitan prefect
Ch'ang-lu salt controller	Chang Chen-fang	Hunan judicial commissioner
Ch'ang-lu salt controller	Liu Chung-lin	Hunan judicial commissioner

Source: *CSL:KH*, *CSL:HT*. For individual references, see below, note 47.

before becoming first deputy in the department of salt administration (*yen-cheng yüan yen-cheng ch'eng*) and subsequently became salt commissioner in Szechwan.[48] Chou and Yang were already men partly outside the tradition, having found their careers in the new openings of the self-strengthening program. K'o and Yen, however, had no such associations. It may be that there was a tradition of men in the salt administration who, realizing that they were unlikely to reach the top echelons of the bureaucracy, decided to become fiscal experts instead and carve out careers for themselves at an ancillary level.

Political involvement. The nineteenth-century Chinese official was frequently a politician as well as a civil servant, and the distinction between politics and administration, so important for the concept of a civil service in England, for example, since the time of the 1854 Trevelyan–Northcote report, hardly existed in China.

Some degree of independence from politics is an essential feature of any truly modern bureaucracy. For while on the one hand the increasing complexity of modern society makes a powerful and highly rationalized bureaucracy ever more necessary, at the same time it tends to reduce its relative importance politically as the other component elements of society in turn diversify and achieve independent status. A modern society will have many elites — political, managerial, artistic — each taking the lead in particular fields; thus the fusion of political and administrative functions in a single dominant elite, as in imperial China, may, from this standpoint, be regarded as distinctly premodern. Both functions were, on a modern assessment, distorted by the fusion: Chinese politics were bureaucratized and could not generate the mass participation in politics required for modernization; Chinese administration, on the other hand, was politicized and could not achieve maximum rationalization, because of the political requirements it had to meet.

This may be illustrated in a narrow sphere, if we consider the development of salt taxation at Ichang, where Szechwan salt entered Huai-nan division (Table 8). Until the 1850's, only the extreme southwestern prefecture of Hupei consumed Szechwan salt, the rest of the province belonging to Huai-nan. With the disruption of the Yangtze trade by the Taiping rebellion, there was a salt famine in Hupei and Szechwan salt flooded in to meet it.

Between 1854 and 1861 taxes amounting to 11.5 cash a catty were placed on Szechwan salt passing Ichang; these were collected by a special agency of the Hu-kuang authorities, the Szechwan likin office (Ch'uan-yen tsung-chü), and were used to finance Tseng Kuo-fan's forces besieging Anking.

Table 8. Ichang revenue collections, 1900.

Tax	Date	Rate per catty (cash)	Amount (taels)	Office allocated to
Regular tax (*cheng-k'o*)	1861	11.5	862,500	Salt likin collectorate of the Maritime Customs
Expenses (*kung-fei*)	1861	1.5	112,500	Szechwan likin office (Ch'uan-yen tsung-chü)
Additional tax (*chia-shui*)	1864	5	375,000	Salt likin collectorate and Szechwan likin office
River defense tax (*chiang-fang*)	1884	2	150,000	Szechwan likin office
Additional likin (*chia-li*)	1884	3	225,000	Liang-kiang additional likin office (Liang-kiang chia-li chü)
Army levy (*ch'ou-hsiang*)	1894	2	150,000	Szechwan likin office
New army levy (*hsin-hsiang*)	1898	2	150,000	New army levy office (lien-ping hsin-hsiang chü)

Source: Customs Report, pp. 42–50. Amounts have been calculated on the basis of a Szechwan export of 1,000,000 piculs of salt (Customs Report, p. 78). Total appropriation by the four offices were as follows: Salt likin collectorate, 1,050,000 taels; Ch'uan-yen tsung-chü, 600,000 taels; Liang-kiang chia-li chü, 225,000 taels; *lien-ping hsin-hsiang chü*, 150,000 taels; grand total, 2,025,000 taels.

With the ending of the rebellion, however, Tseng, as Liang-kiang governor-general, had to consider the interests of his own Huai salt, which wanted its monopoly in Hupei restored; this was all the more important since the taxes collected at Ichang by Hu-kuang would no longer be so readily available to Tseng, as in the dark days of 1861. He insisted, therefore, on the revival of complete Huai-nan monopoly in eastern Hupei and, in the western half where the two salts would compete, on the imposition in 1864 of an additional tax (*chia-shui*) of 5 cash a catty on Szechwan salt passing Ichang, to give the Huai product additional protection. Further, half the proceeds of this tax were to be handed over to the Liang-kiang authorities by the Szechwan likin office which collected it, to compensate for the loss of revenue owing to the decline in sales of Huai salt.

The Hu-kuang authorities, on the other hand, had every reason to encourage the sale of Ch'uan salt since, despite the payment of half the additional tax to Nanking, they obtained nearly twice the revenue from Ch'uan salt as they did from Huai.[49] Thanks to the protection of these authorities, coupled with its own superior quality and Ting Pao-chen's reform of the Szechwan salt administration in the years 1877–1882, Ch'uan salt continued to expand its sales in Hupei, despite two new taxes imposed for revenue purposes by Hu-kuang in 1884 and 1894, a river defense tax (*chiang-fang*) and an army levy (*ch'ou-hsiang*).

In 1884, the Liang-kiang authorities now headed by Tseng Kuo-ch'üan made a new move. They obtained the imposition on all Ch'uan salt passing Ichang of an additional likin (*chia-li*) of 3 cash a catty. This time, however, it was to be collected by a special Liang-kiang agency at Ichang, the Liang-kiang additional likin office (*chia-li chü*). In other words, Nanking was now reconciled to the existence of the Ch'uan salt trade to Hupei and was concerned with obtaining its fair share of the tax. So concerned was the governor-general with this that, to prevent collusion between the Szechwan likin office and the new additional likin office, a special official was stationed at Wan-hu-t'o to report to him directly the amount of salt coming down the gorges.

In 1898 the position was further complicated by the Anglo-German loan agreement, which laid a quota of a million taels on the Ichang revenues, this sum coming from the whole of the regular tax of 11.5 cash a catty and half of the additional tax (*chia-shui*). This involved a serious loss of revenue for Hu-kuang, so in 1899 Chang Chih-tung added a new tax for military purposes of 2 cash a catty, to be collected by an entirely new agency, the new army levy office (*lien-ping hsin-hsiang chü*), which was more under his personal control. By 1900, thanks to these political crosscurrents, the administration at Ichang had reached a highly undesirable degree of complexity: it had the heaviest tax rate anywhere in China, and four collecting agencies in a single city. So long as the salt administration remained intertwined with local politics, this kind of situation was bound to occur, rationality of organization being sacrificed to political exigency.

Ideology: Clichés and Placebos

For all its longevity, the theoretical debate on salt was an infertile, uncreative corpus of thought and this for two reasons. In the first place, it was insufficiently analytic, in the sense that it failed to enunciate and clarify its own presuppositions and aims. A body of unexpressed and hence uncriticized major premises lay behind every Chinese argument on salt affairs: for example, the premise that taxes were not to be consolidated in a single national treasury. Second, the language of the salt tradition was insufficiently empirical: it was a bureaucratic, protocol-oriented officialese, which obscured facts as much as it revealed them. Lack of analytic rigor and insufficient empirical reference — two prime requirements of modern thinking on any subject — created a body of thought too self-contained, too cut off from new ideas and inconvenient facts, for the needs of the modernization process.

This solipsism of traditional Chinese thinking about salt may be illustrated by a 1901 memorial from Yüan Shih-k'ai, governor of Shantung, describing his reforms there:[50] "In obedience to orders, I am managing Tung province salt affairs and have drawn up five regulations for reform: first, to demarcate the *yin* and the *p'iao* areas; second, to define rigorously the salt office jurisdictions; third, to reform the old regulations for transportation and retail; fourthly, to establish increased prices and increased taxation; and fifthly, to eliminate entrenched corruption."

That Yüan was taking vigorous measures of some kind is apparent, but the exact sense of his reforms is concealed behind the formalized account. What is clear is the decidedly traditional framework. The aims of the reform require no mention at all; the *yin* and *p'iao* systems are taken for granted and only require fresh definition; and "entrenched corruption," this unanalyzed disease so frequently mentioned by salt memorialists, is given no causal diagnosis. Yüan Shih-k'ai was among the least traditional of the officials of the last decade of the empire, yet conceptually this memorial belongs to the old order. Even if, as is probable, some of its originality and force have been weakened by the condensation necessary for inclusion in the *shih-lu,* this itself is significant, for such condensation was part of the normal administrative process: memorials, for example, were frequently quoted in imperial edicts (*yü*) in a condensed form. Thus the administrative process itself automatically weeded out any language which did not cohere with its own solipsistic universe of discourse. A traditional officialese and a traditional administration reinforced each other, and together they inhibited any extratraditional changes.

That Yüan's reform did not amount to much is confirmed if one looks at the actual situation in Shantung following his governorship, an exercise which will, at the same time, serve to recapitulate the main points of this chapter.

Shantung salt division,[51] which covered the one hundred and seven hsien of the province itself as well as nine Honan, five Kiangsu, and two Anhwei hsien, was an excellent example of the defects of the premodern salt administration. Only 2,125,325 piculs of salt passed under official cognizance and paid tax, although the division's population of 35 million (a conservative estimate) probably consumed at least 3 million piculs.[52] Over two-thirds of the tax came from the sixty-six *yin* districts supplied by water via the Hsiao-ch'ing River and the Grand Canal: in the fifty-seven *p'iao,* or land-supplied districts,[53] effective rates of tax were much lower, particularly in the eighteen districts of the Shantung promontory, which exported large quantities of salt to Korea. Taxes for both areas, *yin* and *p'iao,* assessed under from twenty-three to forty-three heads, were paid at Tsinan, a hundred miles from the sea, and as there was no effective control of the trade at the salines, smuggling was widespread. At Tsinan itself, the administration was lax: Dane found that bags of salt supposed to weigh 325 catties (and taxed accordingly) were discovered to weigh between 425 and 472 catties, and the salt office did not even possess a proper weighing scale. Organizationally, the salt division lacked uniformity: "official transport and official sales" (*kuan-yün kuan-hsiao*) was in force in twenty-six districts, "official supervision and merchant sales" (*kuan-tu shang-hsiao*) in seventy-nine, and free trade (*min-yün min-hsiao,* lit., "popular transport and popular sales") in the eighteen districts of the promontory.

Yüan's reforms remained well within the traditional compass: all the major structural components of the traditional administration were still in place afterwards. This is not to belittle Yüan Shih-k'ai, but only to underline that the tradition of reform, of which he was here the exponent, was a limited one, both in its concepts and practical effects, and that in 1900 the process described as modernization was not yet envisaged. In its thinking as well as its institutions, the salt administration in 1900 was thoroughly premodern.

2 | Change in the Salt Administration, 1900–1911

This traditional structure of the salt administration as outlined above survived in its essentials until the reforms of the foreign inspectorate in the period 1913–1918. Nonetheless, Dane in 1913 was confronted with a situation significantly different from that described to Hart by the customs commissioners in 1901. The late Ch'ing reform movement, followed by the revolution of 1911, introduced changes in the salt administration which both established the agenda of modernization and undermined the status quo. Consequently, Dane inherited a situation which was no longer static and where the direction of further reform was already indicated.

This chapter examines the changes in the salt administration which resulted from the late Ch'ing reform movement in the decade between the Boxer protocol and the Wuchang uprising. Three new and nontraditional developments, already prefiguring later changes, need examination here: the increasing use of salt revenues for nontraditional purposes, in particular railway finance; the effect of new forms of transportation on the direction of the salt trade; and finally, the development of new institutions and policies designed to make the salt administration more national and comprehensive.

These departures from tradition were not unrelated. They had as their common origin a peculiar coincidence of China's economic and political circumstances, which enhanced the political importance of salt and focused the attention of Chinese officials on its problems and potentialities.

CHAPTER 2

Salt and the Politico-Economic Context, 1900–1911

The salt monopoly in China in the last ten years of the Ch'ing cannot be separated from the life of the empire as a whole. It functioned as part of a complex politico-economic system, and if either the economic or political situation changed, the salt monopoly was bound to change with it. In fact, in this decade, both the economic structure and the political situation in China were changing, and changing materially, though in opposite directions. In the economy, there was a general expansion of activity, while in politics there was growing financial embarrassment for the central government and a widening gap between revenue and expenditure at all governmental levels. It was a situation of "rich country, poor government" whose effect on the salt administration was twofold: economic expansion created the possibility of larger salt revenues; fiscal embarrassment created the need for them. The result was a marked expansion of salt revenue on the one hand and, on the other, a diversification of the uses to which it was put. This diversification of its uses was accompanied by commercial and institutional changes.

An attempt to plot the course of the economy over a decade must be tentative, given the area involved and the premodern character of the statistics. So many local factors — the Russo-Japanese war of 1904–1905, the opium-suppression campaign in Szechwan from 1907 to 1909, the middle Yangtze floods of 1908–1909, the Manchurian plague epidemic of 1910–1911, or the Shanghai rubber speculation of 1910 — enter into the situation that the value of generalization is necessarily limited.[1] With these qualifications, however, an overall impression can be formed from the decennial reports of the Maritime Customs which controlled not only China's foreign trade in a narrow sense, but also the greater part of her coastal and inland navigational trade, since the greater part of this trade was carried in steamships whose supervision had come to be the function of the customs officials.

The impression left by the customs reports is one of general expansion. As the Hankow commissioner put it, "if the history of this decade had to be summed up in one word, 'boom' only would be appropriate." [2] In these years, 1900–1910, China's strictly foreign trade grew from £59 million to £113 million, while coastal trade between treaty ports rose from 268 million taels carried in 51 million tons of shipping in 1905, to 328 million taels carried in 63 million tons in 1910.[3] Prosperity was not evenly spread: the boom was more marked at Hankow, Tientsin, and the Manchurian ports than at Canton or Shanghai, the main reason for this being that it was based on a combination of Chinese railway development and world demand for sesamum (Hankow), raw cotton (Tientsin), and the soya bean (Newchwang and Dairen). It was, that is, an export and railway boom, rather than a simple, overall expansion on the old basis.

Conditions were excellent for buoyant salt sales. The region of maximum prosperity coincided with the principal traditional market for salt. The exports which swelled Hankow's trade from 73 million taels in 1902 to 135 million taels in 1910, and Tientsin's from 71 million in 1904 to 116 million in 1911, came not from the immediate environs of those cities but from the remoter hinterland in the interior. At Hankow, the commissioner of customs wrote: "the trade feature of the decade has been the growth of sesamum seed . . . it has made Honan and the Hankow hinterland rich." At Tientsin "The outstanding feature of the export trade is the great advance made in the shipments of raw cotton," and this the commissioner put down to improved railway facilities, especially with western Chihli, Honan, and Shantung.[4] These developments enriched inland provinces such as Honan and Hupei, which were part of the central Chinese market for salt, the principal export outlet for the three chief salt divisions — Ch'ang-lu, Liang-huai, and Szechwan.

In areas where an increase in exports was not much in evidence, a rise in imports sometimes stimulated the salt trade. For example, Szechwan in this decade had an adverse balance of payments with the outside world. Imports rose, but poor communications made for sluggish exports, a situation exacerbated by the ending of the opium trade, her principal outside exchange winner. In 1914, and the customs reports indicate similar monetary conditions earlier, the import houses at Chungking were reported to be expanding sales of salt to Hupei, to obtain foreign exchange for their imports from Hankow. It is not surprising therefore to find a memorial in the *shih-lu* reporting the invasion of the Huai salt area by Ch'uan salt, or that by 1911 Szechwan had surpassed Liang-huai as the premier salt producer of the empire.

Finally, the salt trade was beneficially affected by the marked increase in prices, from between 50 and 100 percent generally in all areas in China. At Chefoo, Shantung salt doubled in price over the decade; at Chinkiang, the price of the salt catty rose from 3 cents in 1902 to 7 cents in 1911; at Wuchow, the corresponding figures were 4 cents and 8 cents; and even in distant Tengyueh on the Burmese frontier salt prices went up from 2 cash an ounce to 6 or 7.[5] To some extent this increase was due to higher taxation — this is specifically mentioned at Lappa where the price quadrupled — but since the rise in salt prices was markedly in line with the increase in other commodity prices (at Wuhu, the chief rice-exporting center for Anhwei, the picul price rose from $1.80 in 1904 to $2.80 in 1911), it may be put down chiefly to the general inflation and increasing prosperity. Economic conditions in the prerevolutionary decade thus favored the salt trade, and by the same token increased its taxable capacity.

The boom made an increase in the salt tax feasible; the politico-fiscal situation made it necessary. For the paradox of China at this time was that while the country in general was growing richer, the state or, more exactly, the amalgam of central, regional, and local power blocs which formed the political structure, was facing increasing financial difficulties.

Since the Taiping rebellion, political power had been financed by the Maritime and native customs, the salt tax, likin, and provincial land taxes, in that order of importance. From the first Sino-Japanese war onwards, especially after 1900, this tax basis began to disintegrate. The customs had to be hypothecated to meet the Japanese and Boxer indemnities; the eventual abolition of likin had been accepted by the Chinese authorities under the Mackay treaty of 1902; while the land tax quotas were inelastic and the prestige of the dynasty was involved in their maintenance at current rates. All this took place in a decade when prices were rising rapidly and when expenditure on reform projects, such as the new army, railways, or the new educational system, mounted every year.

Further, the very reforms carried out by the imperial government and the governors-general tended to reduce their revenues. Thus Ts'en Ch'un-hsüan's abolition of the lottery at Canton cost Liang-kuang annually 4 million taels; Chao Erh-hsün's vigorous implementation of the opium-suppression campaign caused serious financial difficulties in Szechwan; while his brother Chao Erh-feng's reassertion of Chinese authority in Tibet disrupted the tea trade from which the province had derived a substantial revenue.[6]

The effect of this drying up of so many sources of revenue, combined with the marked inflation and rising government expenditure, was to increase the importance of salt as the only major tax source both comparatively unimpaired by hypothecation and still capable of further expansion.[7] When the Canton lottery and the Szechwan opium tax were abolished, in both cases salt taxes were increased to substitute for them.[8] Indeed, "my solution is an increase in the price of salt" [9] became a frequent refrain in the memorials of harassed governors-general, and the salt tax came to seem a panacea both to them and to the court itself.

Ch'ang-lu salt division is an example both of the motives for the increases and of their scale. Between 1900 and 1911, thirteen new taxes were laid on the Ch'ang-lu salt trade. They included items for the Boxer indemnity, opium tax substitution, railway building, school expenses, industrial development, and army costs, a cross section of late Ch'ing activities, and put together, they raised the yield of the salt taxes in Ch'ang-lu from 1,800,000 taels in 1900 to 5,236,800 taels in 1911.[10]

The total increase for the whole country is difficult to determine exactly, partly because not all increases are recorded in the *shih-lu* and partly because of the uncertainties surrounding gross, reported, and available tax collections referred to in the last chapter. A memorial from the Grand Secretariat in October 1911 stated that the salt taxes levied in the empire had risen from 29 million taels in the last year of Kuang-hsü to 40 million taels in the current year, and it may be estimated that between 1900 and the end of the empire, gross salt taxes rose from 24 million taels to between 40 and 50 million taels. Certainly the court was not exaggerating when, in an edict to the Grand Council not long after Li Hung-chang had memorialized the amount of the Boxer indemnity, it declared: "salt affairs are the staple of the state's annual income." [11]

New Uses for the Salt Revenues

To what purposes were the increased salt revenues applied? Inasmuch as the tax receipts were not consolidated into a single account, but since instead each tax source was earmarked, even before collection, for specified items of expenditure, the *shih-lu* contain numerous entries recording the apportionment of salt revenues to defray particularized expenses.

Some of these were of a traditional character: for example, not long before the Boxer rebellion, an imperial rescript, issued in response to a memorial from Chang Chih-tung, assigned the cost of moving 30,000 *shih* of Hunan rice to the capital to the Ichang salt likin receipts. Again, there are two instances of a routine memorial from the board of revenue proposing to assign 600,000 taels from the salt and tea taxes for the expense of the imperial household department.

More significant however than these traditional uses were the newer calls made on this revenue. The earliest was the Boxer indemnity (*hsin-an p'ei-k'uan*). Increases in the price of salt in connection with this levy are mentioned for Honan, Kweichow, Hukuang, Liang-kuang, and Liang-kiang.[12] The memorial of the Honan governor Chang Jen-chün though brief is not untypical: "I have investigated the receipts from the Lu salt *hsin-an* increase of price. These are to be specially devoted to the needs of the indemnity."

Another application of salt revenues connected with the events of 1900 is found in a memorial from the Mukden tartar-general Tseng Ch'i, proposing to reform the Manchurian salt laws in order to provide for the Russian army of occupation. Military modernization, too, drew financial support from the salt revenues. For example, a memorial of Shantung governor Chou Fu, reporting a deficit in the funds for the *wu-wei yu-chün* (lit., "right division of the guards army"), the nucleus of Yüan Shih-k'ai's army, requested an increase in salt price to make up for it, and a little later the same official, now Liang-kiang governor-general, proposed an increase in salt likin to pay for a Japanese-built, shallow-water fast gunboat.

Yet another field of activity in which the salt revenues played a part was education. The ending of the traditional examination system gave an impetus to the building of schools. In Hunan, perhaps the most advanced region educationally in the empire, the governor memorialized that the school superintendent wanted to raise the price of the Liang-huai salt catty one cash in order to meet expenses.

But the most important of the new uses for salt revenue was in railway construction. Railway building, so prominent in the decade before the revolution, served a dual function. On the one hand it directly fostered economic expansion by lowering transportation costs and opening new markets. On the other hand, less obvious perhaps, it had the function of siphoning off some of the new wealth into the financially embarrassed political system, railways at this time being a source of both money and influence. The Peking to Mukden railway for example, opened in 1903, served both to extend Chinese authority in Manchuria and to finance the Peking to Kalgan line which, officially opening in 1909, in turn strengthened the Chinese hold on Inner Mongolia. All levels of the power structure — the court, the governors-general, the provincial gentry, and the merchants — turned to railway projects as a means of grafting themselves onto the expanding economy and of avoiding the general disintegration of political power being caused by the drying up of formerly reliable sources of revenue.

Railway building, however, was expensive, and, as the only major source of revenue not mortgaged to foreign bondholders, the salt monopoly was a major source of funds, being called upon to contribute either directly to the capital cost, as with the proposed Hankow–Canton line in 1908, or indirectly, by the servicing of foreign railway loans. The loan might be specifically secured on the provincial salt revenues; this was the case with the consortium Hu-kuang loan of 1911.[13] Alternatively, interest might simply de facto be raised by means of increased salt tax as in the case of the Tientsin to Pukow line. Another way of using salt revenues was in dividend payments; for example, in Shansi the governor recommended that the price of salt be increased so that the railway company could raise its dividend from 4 to 8 percent.

Altogether there are some twenty decrees and memorials in the *shih-lu* dealing with the use of salt revenue for railway building, covering eight railway projects: Hankow–Canton, Hankow–

Szechwan, Tientsin–Pukow, Chin-chou–Aigun, the Shansi and Lunghai railways, and lines in Kiangsi and Inner Mongolia. They suggest two conclusions.

In the first place, the initiative seems to have come as much from local interests as from the governors-general who officially sponsored the lines. For example Yang Shih-hsiang, Chihli governor-general memorializing on behalf of the Tientsin–Pukow line, stated that the gentry had petitioned that the price of Ch'ang-lu salt in Chihli be raised 4 cash a catty to meet railway expenses. This gentry initiative may be seen as part of that development of local interests independent of, though not necessarily opposed to, the viceregal yamens, which was characteristic of the politics of the late empire. The gentry's reliance on salt for railway finance was partly prompted by the court's similar mobilization of resources for the Boxer indemnity. A memorial of Hu-kuang governor-general Ch'en K'uei-lung stated: "The requirements of the Yüeh–Han railway are urgent and critical. Gentry and people have consulted together and according to the precedent of increasing the salt catty contribution for the Boxer indemnity, request an increased contribution of 4 cash a catty on all salt, Ch'uan, Huai or Yüeh, sold within the boundaries of Hunan." [14]

In the second place, it is clear that these local initiatives were not always successful, failure being due to conflicts between the railway construction areas with a primary interest in taxing salt and the wider interests of the salt division in which they were situated, which had a primary interest in selling it. For example, Kiangsi province had ambitious plans to build a line from Kiukiang via Nanchang to link up with the Hankow–Canton line at Shiuchow in northern Kwangtung, thus replacing the old route of the ambassadors along the Kan River, and wished to raise the salt catty price 4 cash to finance it. The head of the Liang-huai salt division, Liang-kiang governor-general Tuan-fang, originally supported this proposal, but his successor Chou Fu was less enthusiastic: "Kiangsi because of undertaking the railway has increased the price of Huai salt. This has obstructed the sale of the quota and I request immediate cessation." [15] The court sided with the governor-general and no railway was built.

A similar situation occurred with Ho-tung salt in Shensi: the Shensi authorities had increased the price of salt to pay for a railway, presumably their section of the Lunghai line, and the governor of Shansi, the head of the Ho-tung salt division, protested that this was affecting the sale of his salt. Control and taxation of salt itself thus became an issue separating local interests from the yamens of governors and governors-general, and this occurred whether or not the proposed railway was in fact built.

What proportion of the salt revenue was devoted to railway building, and how did this stand in relation to the total expenditure on railways? In Ch'ang-lu in 1913, salt taxes in Chihli province came to 8.757 taels per *yin,* with the Tientsin–Pukow railway item of 1.600 taels per *yin* first levied in 1909 being the largest single item. In Honan, where there was a similar contribution for the Lunghai railway, the figures were 9.060 taels per *yin* for total salt taxes and 1.620 taels for railway expenses.[16] A significant percentage of the Ch'ang-lu salt revenues was thus being devoted to railway purposes.

In relation to the total capital cost of the railways, the special salt contributions, it is true, were small. The Chihli item just mentioned would have raised annually a mere 750,000 taels, while the Tientsin–Pukow railway involved foreign loans of £9,800,000 or 65 million taels. On the other hand, 750,000 taels would amount to nearly a quarter of the 5 percent interest on the loans, and this was probably the immediate concern of the authorities.

In Hunan, according to a memorial of governor Yang Wen-ting setting forth the provincial budget for the third year of Hsüan-t'ung, total provincial expenditure of all kinds came to 9,232,700 taels, of which 915,700 taels were marked for railway expenses.[17] This was met by two items, good harvest tax and salt tax which together produced 916,000 taels. What proportion of this came from the salt? Hunan was at this time the largest single importer of Huai salt and its quota stood at over a million piculs a year. It also imported a quantity of Ch'uan and Yüeh salt which may be estimated at another million piculs. Governor-general Ch'en K'uei-lung's proposal of a general Hunan tax of 4 cash a catty would thus have brought in 666,000 taels, two-thirds of the province's total expenditure on railways. Hunan furthermore was fortunate in having a readily taxable commodity such as rice: in other provinces, such as Chihli, the proportion of railway expenses laid on the salt revenue may have been higher still. A substantial percentage of immediate railway expenditure was thus being met by increased salt taxes in the prerevolutionary decade; though not the only, this is perhaps the most striking instance of the redirection of this traditional source.

Modern Transportation and the Salt Trade

Between 1901 and 1912, eighteen new railway trunk lines were opened for use in China, totaling over five thousand miles, as set out in Table 9. At the same time, the number of steamships

Table 9. Railway construction in China, 1901–1912.

Date of completion	Name or termini	Mileage
1901	Chinese eastern railway	921
1903	South Manchuria railway	591
1903	Peking–Mukden	525
1903	Taokow–Tsinghwa	94
1904	Tsingtao–Tsinan	256
1905	Peking–Hankow	755
1906	Swatow–Ch'ao-chou	26
1907	Chengting–Taiyuan	151
1908	Shanghai–Nanking	193
1909	Peking–Kalgan	125
1909	Shanghai–Hangchow	118
1909	Kaifeng–Loyang	115
1910	Laokay–Kunming	289
1911	Mukden–Antung	189
1911	Canton–Kowloon	90
1911	Amoy–Kiangtungkiao	19
1912	Ch'ang-ch'un–Kirin	79
1912	Tientsin–Pukow	629

Source: Decennial Reports 1902–1911; Chang Kia-ngau, *China's Struggle for Railroad Development.*

owned and operated by Chinese, and thus operating mainly in domestic trade, increased from 570 in 1902 to 901 in 1911.[18] In the long run, the balance of China's communications system was altered by this development of modern transportation. A striking feature of imperial China in the nineteenth century had been the superiority of east-west communications over those running north and south. The Grand Canal had never been the equal of the Yangtze as a means of communication, and from the mid-nineteenth century onwards, it ceased to function effectively at all. This decline in the north-south routes was offset by the development of coastal steamship traffic from Tientsin and Newchwang to Shanghai and the Yangtze ports, but the balance was not fully redressed until the building of the railways. From then on, there took place a general reorientation of China along a north-south axis, increasing the national importance of Manchuria, Peking, and the communications crossroads of Wuhan.

How did these new routes affect the structure and direction of the salt trade? Their long-term effect was to enhance the importance of northern salt at the expense of the traditional Yangtze network, particularly within Huai-nan salt division. By 1922 most of the salt consumed in Huai-nan came from outside the salt division, and by 1955 85 percent of China's edible salt came from the "four big salines" of Liaoning, Ch'ang-lu, Shantung, and Huai-pei.[19] The rise of this northeast arc was already prefigured in the decade before the revolution.

First, in Manchuria the impact of modern transportation was conspicuous. When Sir Alexander Hosie, consul at Newchwang from November 1894 to July 1897 and from April 1899 to April 1900, made his survey, and again when the customs report was written in 1901, the salt administration of the three eastern provinces was still in a primitive condition, although Manchuria was already one of the largest producers. Control over the upcountry trade to Kirin and Heilungkiang was particularly limited: "Until the year 1908 no official cognisance — apart from the collection of likin on salt, as on all other goods, in transit — was taken of the salt trade in these provinces." [20] It was the growth of the railways and of modern port facilities at Vladivostok which made possible the development before 1911 of a salt administration which was the fourth largest in China in terms of salt handled and which Chang Chien regarded as the most efficient in the empire.

The new organization, the work of the first governor-general Hsü Shih-ch'ang and of the head of the Kirin–Heilungkiang transportation office Lu Tsung-yü, was closely linked to the new transportation. In the north, "practically all the salt consumed in Heilungkiang was sent by sea to Vladivostock and from there by rail to Harbin," except in winter when the port was closed by ice; then it was sent directly by rail from the Fengtien coast, where it was produced, to Ch'ang-ch'un and Harbin. In Kirin, "Rather more than two-thirds of the total quantity of salt consumed in the province was sent from Newchwang by the South Manchurian Railway." The traditional carts and river boats, so vividly described by Hosie, were a thing of the past, and Manchurian salt had moved onto a modern communications basis.[21]

Only slightly less strongly affected than Manchuria by the railway were Ch'ang-lu and its close associate in salt affairs, Shantung.[22] Here the shift to modern transport was accompanied by an expansion of sales area. In Ch'ang-lu, the salt merchants were abandoning the half silted-up Wei River, the traditional route to Honan from Tientsin, in favor of the Peking–Hankow railway; and in Shantung, at the main distributing center of Huang-t'ai-ch'iao just outside Tsinan, "the bulk of the salt . . . [was] carried by the Tientsin–Pukow Railway," and "salt for Honan, Kiangsu and Anhui was transported by rail and cart to the district of consumption." [23]

Besides using the railway to service the trade of its own consumer areas, Ch'ang-lu salt was pushing inland beyond these borders. In neighboring Ho-tung salt division, "about 60 per cent. of the quantity formerly sold in the boundary districts has been lost to Hotung District, as the Changlu merchants, with the advantage of cheaper transport by rail, undersell the Hotung merchants." Similarly, further south, the two Jukuang prefectures of Honan "originally formed part of the Yinti area for Huaipei salt, but after the construction of the Peking–Hankow railway line Changlu salt began to be transported to them." [24] Again, in the north, according to the Chahar commander in chief Ch'eng-hsün, the Mongol princes of the Silingol league were deeply disturbed that the newly completed Peking–Kalgan railway would bring an influx of Ch'ang-lu salt which would obstruct the sale of their own Mongolian salt, on which their incomes to some extent depended — an illustration of Owen Lattimore's theory of secondary imperialism in these years.[25]

Finally, besides these local expansions, northern salt generally was beginning to invade the markets of the Yangtze valley and districts further south by means of steamer transport. In 1908,

Feng Ju-k'uei, governor of Chekiang, itself an important salt-producing province, requested permission to bring in Ch'ang-lu salt by sea. A little later, Liang-kiang governor-general Tuan-fang memorialized that Huai salt production was insufficient for the demand and requested the importation by steamer of 100,000 *yin* of Ch'ang-lu salt and 60,000 *yin* of Shantung salt. Similarly, early in the Hsüan-t'ung period, Liang-kuang governor-general Chang Jen-chün asked to import 200,000 piculs of Ch'ang-lu and Shantung salt by steamer because "Yüeh province has had much rain and fog, the salt has not been able to be sunned and the price has risen to match the scarcity." [26] In 1913, the four Yangtze provinces of Hupei, Hunan, Kiangsi, and Anhwei reported that, in addition to the Huai-nan salt which traditionally supplied the region, "Large quantities of salt are also shipped direct by steamers, Fengtien and Shantung salt coming to Hupeh, Changlu salt to Yochow in Hunan, and Huaipei salt to Hupeh, Kiangsi and Anhui." Dane commented on the reasons for this: "The ostensible reason for this 'borrowing' was a deficiency in the supply of Huai salt. The real reasons were the introduction of the use of steamers for the transportation of salt, and the greater cheapness of the salt made in the North." [27] Northern salt being *shai,* or solar-evaporated, was anyway cheaper than the artificially boiled Huai-nan salt, and, furthermore, being carried by steamer and thus being subject to the Maritime Customs, it could claim exemption from likin en route. The steamship therefore made it a real competitor in the south against the traditional sources of supply.

Thus, with the opening of new means of communication, the traditional pattern of the salt trade began to change. The pre-eminence of the east-west axis along the Yangtze, which had been characteristic of the salt trade since the days of Marco Polo, was now challenged by the establishment of an improved north-south axis. This development involved more than a regional shift within the ancien régime: while primarily benefiting the northeast salines, Fengtien, Ch'ang-lu, Shantung, and Huai-pei, the new routes brought a modern-style dynamism into the slow-moving world of the traditional salt trade. Changes in transportation, a new and exogenous factor, imparted an impetus to modernization in the salt industry.

New Policies and Institutions

If the commercial pattern of the salt monopoly was changing in the first decade of the twentieth century, so too was the institutional pattern. The increased political importance of the salt revenues led to increased competition for their control between the different levels of the Chinese power structure, and this led to projects for institutional change, involving the transfer of control from one political level to another. It must be emphasized that these projects were not successful at the time, but they represent, in effect, the first steps taken by the Chinese in the direction that was later followed by the foreign inspectorate.

Competition in China within the power structure at this date took two forms. On the one hand, there was tension between the court and the regional power blocs of the provincial governors-general. This was a long-standing rivalry, well regulated and institutionalized, kept within bounds by the common interests of both parties. On the other hand, there was a newer, and hence less well regulated, conflict between the governors-general and the rising provincial particularism which expressed itself locally through the provincial assemblies and bodies such as the Szechwan railway protection league, and nationally through the national assembly and, more surprisingly, through the censorate.[28]

The effect of these two tensions on the salt administration was itself twofold: the rivalry *within* the provinces between governors-general and local interests was expressed in alternative plans for the rationalization of the local administrative machine of the salt monopoly; the rivalry *between* the provinces and the center led to the creation for the first time of a centralized, national salt administration. In the salt administration — as in local government, budgeting procedure, railway administration, the armed forces, and the Maritime Customs — the last decade of the empire saw the initiation of fundamental institutional changes.

Administrative rationalization. Administrative rationalization in the period 1900–1911 took two forms. The first of these was the "official transport" (*kuan-yün*) movement, which was designed to extend direct official participation in the monopoly. "Official transport," that is, the purchase and transportation of salt by officials, was a traditional weapon in the armory of Chinese salt administrators; but, in the past, it had been utilized only in exceptional circumstances, for example in backward Yunnan in the eighteenth century, since in general the initial capital cost of buying the salt was beyond the resources of the government, and, furthermore, the system involved an undesirable contamination of officials by business. By 1913, however, *kuan-yün* had become the official policy of the gabelle; as Dane stated, "A Government monopoly was supposed by all Chinese experts to be the ideal system of administration." [29]

Converging factors lay behind this change of attitude. Administratively there was the successful operation of the "official transport" system in Szechwan where it had been introduced by Ting Pao-chen, governor-general from 1876 to 1886. By 1911, Szechwan had surpassed Liang-huai as the premier salt-producing area of the empire and this reflected credit on the system used by Ting. Thus an edict of 1901, handed down just after Li Hung-chang had announced the size of the Boxer indemnity and ordering the governors-general to deliberate how to meet it, commended Ting's example: "Formerly as regards Szechwan salt affairs Ting Pao-chen broke down prejudice and carried out genuine reforms, so that down to the present, the increase of the quota has been very great. We ought in like manner to organize in the hope of increasing the treasury quota." [30]

With respect to communications, railway development favored "official transport" since, being in general the cheapest form of transport, it could compete against all other forms of transport and it was already under official control. The Kirin–Heilungkiang monopoly set up in 1908 is the best example of railway "official transport." Even Dane who disliked the system was forced to admit that "The circumstances of the two Provinces are particularly favourable for the successful working of a monopoly." [31]

Theoretically, the example of foreign countries worked in the same direction. In the period under review, the court received reports on the salt administrations of Japan and Italy, both of which, in common with many European countries at the time, operated full government monopolies.[32] "Official transport" therefore seemed the most modern system of salt administration.

Proposals for the introduction of "official transport" were made in several areas in the period 1900–1911: Kwangtung, northern Shansi, Jehol, Manchuria.[33] The most significant scheme, however, because of its modernizing character, was that of Chao Erh-hsün, governor-general of Szechwan in 1909.

In accordance with the basic principle of the traditional salt administration, Ting Pao-chen's system of "official transport" extended only to waterborne salt, called in Szechwan *yin* salt. Landborne or *p'iao* salt, which supplied nearly half the province, was left to private enterprise and went almost untaxed. Chao Erh-hsün, one of the most vigorous officials of the late empire, who had stamped out opium production in Szechwan and helped his brother Chao Erh-feng organize the reconquest of Tibet, planned to end this situation by extending government transportation to landborne salt. As an imperial rescript put it, he wished to "abolish the *p'iao,* [and] reform the *kuan-yün,*" [34] thus substantially increasing the tax yield. In the end, Chao's scheme was never put into effect, being opposed by both the censorate and by local officials,[35] who claimed that official transportation of salt would injure local merchants while proving unable to supply the consumer adequately. Because of its intention to make the salt tax comprehensive, Chao's proposal has a definitely modern character and forms an interesting anticipation of Dane's later attempt to tax landborne salt.

The second form of administrative rationalization planned in the period 1900–1911 consisted of proposals to replace the multiple tax collections of the old regime with a single collection at the salt field, a program expressed by memorialists in the phrase *chiu-ch'ang shou-shui* (lit., "at the yard receive taxes"), or taxation at source.

This was a traditional phrase: taxation at source was the system attributed to Liu Yen of the T'ang; in the seventeenth century it had been advocated by Ku Yen-wu; and in the nineteenth century it had been applied in special cases, such as Yunnan where the concentration of the salt wells in a restricted area and the paucity of the outlets from them made control at source practical. In general, taxation at source was rejected as administratively impractical by traditional salt reformers, such as T'ao Chu, because it required a degree of control over salt fields and trade routes beyond the power of the administration. In the late empire, however, the system as a form of administrative rationalization was becoming popular among officials concerned with salt, the *shih-lu* containing five references to it, in the context of reform.[36]

The most interesting of these references, because of its bearing on Dane's plan, appears in a memorial of Liang-kuang governor-general Ts'en Ch'un-hsüan of June 7, 1907: "salt affairs are the great item to be settled. I propose that we appoint an official to go overseas to investigate the *chiu-ch'ang shou-shui* system and deliberate on copying it in our management." [37]

Ts'en was a notable reformer, one of a group of officials who rose to prominence in the course of the empress-dowager's flight to Sian in 1901. He commanded her escort on that occasion and became successively governor of Shensi and governor-general in Szechwan (1902–1903) and Liang-kuang (1903–1907). In June 1907 he drew up a comprehensive plan for railway development and was appointed minister of communications to put it into effect. He fell suddenly from power later in 1907,[38] and remained out of office till October 1911 when he was reappointed to Szechwan in an effort to conciliate the rebels. He refused office, however, joined the revolution, and later figured as a Kuomintang member and one of the ringleaders in the revolt against Yüan Shih-k'ai in 1916.

The taxation-at-source proposal was made when Ts'en was at the height of his power, and when it seemed that he was becoming a major figure in the late empire. It is no accident that his salt reform project appeared so soon before his railway program: taxation at source would not only increase the revenue, since like Chao Erh-hsün's "official transport" plan it would tax all salt and not merely waterborne salt; but furthermore, by confining taxation to one place, it would lay the way open for the consolidation of the numerous salt levies of the ancien régime into a single tax, which would then be at the disposal of the central government. Taxation at source, if it could be successfully operated, would be both comprehensive and centralized, an excellent base for the ambitious railway project.

Unfortunately, the outcome of Ts'en's proposal for a mission to go overseas is not clear, though proposals for taxation at source continued to be made after his fall. E. H. Parker stated without citing his source that "previous to the death of the Dowager and Emperor in 1908, a Chinese mission had already been sent to India to inquire into the nature of the Salt Administration there," [39] and this may well be Ts'en's mission since India was the principal country to operate the taxation-at-source system. A Chinese delegation was certainly in Simla "about the year 1906," where it was entertained by the foreign secretary to the government of India, Sir Louis Dane. As Sir Richard Dane used to spend much of his leave with his brother, "it is more than likely that he made contact with the Chinese delegation," [40] especially if this was in fact the salt mission referred to by Parker, for Sir Richard Dane was at that time inspector-general of salt and excise to the government of India. Thus it is not impossible that the example of British India, a modern-style salt administration using the taxation-at-source method, was already affecting Chinese thinking about salt, several years before Dane himself applied the Indian system in China.

Centralization. Both the "official transport" and taxation-at-source projects were associated with provincial governors-general. The court, however, no less than other levels of the power structure, was threatened by the disjuncture between the rising cost of government and the financial resources available, and tried to compensate for this by extending its authority over sources of income such as the railways, the Maritime Customs, and, inevitably, the salt monopoly. Though some efforts in this direction can be traced as far back as Kang-i's attempt before the Boxer rebellion to increase remittances to Peking by eliminating corruption, the salt administration remained provincial rather than imperial until the end of 1909.

On December 31 of that year, at the urging of the ministry of finance, the post of controller-general of the salt gabelle (*tu-pan yen-cheng ta-ch'en*) was created, followed on January 19 by the establishment of the office of the controller-general of the salt gabelle (*tu-pan yen-cheng ch'u*) holding general jurisdiction over all salt affairs.[41] Duke Tsai-tse, finance minister and a rising figure in the government, particularly in railway affairs, became controller-general. The first act of the new body was to issue regulations defining its own authority in general terms.

The provincial authorities, however, resisted, the lead being taken by Hsi-liang, the Mongol governor-general of Manchuria, despite the fact that they were to be associated with the new system as associate controllers-general. An imperial edict was eventually issued confirming Duke Tsai-tse's authority, but it is clear that this proved insufficient inasmuch as in 1911 the Grand Secretariat proposed that the powers of the office of the controller-general of the salt gabelle be increased by converting it into a department of salt administration (*yen-cheng yüan*), that it be made, in effect, a full ministry which would operate through its own hierarchy of officials, and not through the provincial authorities except in matters genuinely local. The court accepted this proposal and Duke Tsai-tse became the new minister (*yen-cheng yüan yen-cheng ta-ch'en*). The intention was clearly to create a centralized national salt administration.[42]

Did the new institutions make any practical difference to the conduct of the salt administration? In particular, did they succeed in consolidating all salt revenue into a single account under the effective control of the central government and thus anticipate Dane's most notable reform? As this question is crucial for evaluating the achievements of the foreign inspectorate, the evidence must be considered in detail.

It seems certain that the first of the new institutions, the office of the controller-general of the salt gabelle, did not achieve much in the way of fiscal centralization. In the first place, the memorial which proposed its transformation into the department of salt administration gave as its reasons that the provinces were a law unto themselves (*ko-sheng tzu-wei feng-ch'i*) and incapable of planning for the public benefit (*pu-neng mou kuo-yü-min kung-i*). In addition, the *shih-lu* continued to refer to quotas and specific assignments of salt revenue as before: in the Hunan budget for 1911, the good harvest and salt taxes were specifically assigned to railway expenses, and in Honan, governor Pao Fen reported that the provincial assembly had petitioned for a contribution from the salt for similar purposes.

With the advent of the department of salt administration, whose avowed purpose was to increase central control, the evidence becomes more equivocal. On the one hand, the imperial budgets for 1910, 1911, and 1912 give the aggregate figures — with breakdowns — of the salt revenues of the whole empire, and some contemporary Western observers took this aggregation at its face value. For example, the editor of the *National Review* in his preface to the English version of Chang Chien's *A Plan for the Reform of the*

National Salt Administration, published in 1913, wrote: "Before 1900 the Hu Pu (Board of Finance) received in revenue from the Salt Tax a sum slightly over Tls. 12,000,000 . . . Through the improvements and reforms effected by the Chinese Government and by it alone, the revenue now stands at Tls. 45,000,000 . . . the Chinese Government has more than trebled its revenue from the Salt Gabelle." [43]

On the other hand, the evidence against this interpretation of the late Ch'ing reforms is strong. First, the memorial which prompted the establishment of the department of salt administration was dated October 7, 1911,[44] and it is difficult to believe that a major fiscal reform was effectively accomplished in the few days left before the outbreak of the revolution.

Corroborating this, Dane later gave as his opinion that "The amount which reached the Central Government varied from year to year, but the amount received in any one year does not appear to have ever exceeded Tls. 13,000,000," and he commented on the salt revenue consolidation plans generally that "the Revolution broke out before any definite results were obtained." [45]

Again, although the *shih-lu* contain only a dozen entries on the salt administration between the establishment of the department of salt administration and the end of the dynasty, two of them, memorials from the Hunan authorities and the Chihli governor-general Ch'en K'uei-lung, refer to the system of quotas and assignments in the traditional manner. That from Hunan, for example, requested the court to remit 440,000 taels salt and general likin assigned to a variety of purposes including railways, the army, metropolitan famine relief, and the expenses of Peking university.

Finally, the earliest surveys of the foreign inspectorate show a multiplicity of salt taxes still being levied locally for particular purposes: for river works in Ch'ang-lu, for the Lunghai railway in Ho-tung, for gunboats in Huai-nan.[46]

Fiscally then, it is doubtful if the new institutions made much impression on the old regime.[47] The imperial budgets may best be regarded in Dane's words: "As an estimate of the revenue, which it was considered possible to collect from the taxation of salt under an improved system of administration," [48] rather than as a statement of what was genuinely available to the central government, though this in itself was an impressive administrative achievement.

Indeed, administratively there was a new degree of centralization. Before 1910 suggestions for reform in any part of the salt administration had come always from the provincial authorities involved — from the Liang-kiang governor-general for Huai salt and so forth — and not from the board of revenue. From 1910 onwards, however, the controller-general of the salt gabelle can be found initiating piecemeal reform programs in regions as widely separated as Huai-pei, northern Shansi, Yunnan, and Szechwan. In Fengtien, for example, he suggested the establishment of a salt controllership *(yen-yün shih)* to supervise the production and distribution of salt throughout the whole of Manchuria, a post to which Hsiung Hsi-ling, later prime minister at the time of Dane's reforms, was appointed.[49] It is possible that Duke Tsai-tse's plans were not always put into effect, yet these programs suggest that his grasp of local affairs was firmer than the board of revenue's had been, and that given time, the new institutions would have developed into a genuinely centralized and national salt administration.

Conclusion

Together with most Chinese institutions, the salt administration was changing in the decade before 1911: the revolution was to emerge from a background of growth, growth which in the salt monopoly already anticipated reforms that were to be carried out later under the foreign inspectorate. Among these were the mechanization of transport, the rationalization and standardization of taxing procedures, and the centralization of overall control.

The late Ch'ing reform movement initiated modernization projects in a wide range of institutions; railway nationalization, centralization of the armed forces, the creation of the board of customs control *(shui-wu ch'u)* to end the autonomy of the Maritime Customs — these were reforms of real significance. Yet because of the political weakness of the court grandees and provincial governors-general who initiated them, these projects were not carried through to completion, and the attempt to implement them precipitated the collapse of the political system in which they had originated. Consequently, at the end of the dynasty, the railways remained unnationalized, the armed forces retained their regional loyalties, and the Maritime Customs continued to be an *imperium in imperio*. Political disintegration nullified institutional development.

The changes in the salt administration between 1900 and 1911 fell into this pattern of modernization manqué. Despite new forms of spending, most salt revenues were still expended in traditional ways. Despite the railways and the new north-south axis, the Yangtze was still the main artery of trade and the salt junks still sailed from Shih-erh-wei and Lu-chou. Despite Chao Erh-hsün, Ts'en Ch'un-hsüan, and Duke Tsai-tse, there was no uniform, national system of salt taxation. In sum, despite reform, tradition still ruled.

3 | The Salt Administration in the Revolution of 1911

If tradition still ruled in the salt administration in October 1911, it did not do so for long. Between the outbreak of the revolution and April 1913, when the reorganization loan agreement was signed, a number of new factors produced changes in the gabelle, amounting to a breakdown of its traditional structure. In the first place, the traditional vocabulary and methods of salt reform were rendered outmoded by the rise of a salt reform movement that was modern in argument and organization, and at the same time the old pattern of salt administration was broken by major administrative reforms in seven out of the twelve salt divisions. Next, a complete break with the past was made when, in May 1913, the entire salt administration passed under the control of foreign officials whose offices were established by the reorganization loan agreement.

However, even though the old system was crumbling, the reconstruction, initiated by these new factors, was tentative and inconclusive: the salt reformers could not win the ear of the government; the changes in the salt divisions were contradictory; and the foreign officials seemed strong enough only to thwart the endeavors of the Chinese reformers.

The processes at work in the salt administration between 1911 and 1913 destroyed more than they built up. In this they were only a microcosm of the revolution of 1911 itself. Since one of the new factors operative in the gabelle was the confusion and disorder caused by the revolution, it will be useful to preface this chapter with an analysis of the political episodes of the revolution, from the limited point of view of the history of the salt administration.

Salt and Politics, 1911–1913

The revolution of 1911 may be interpreted not so much as a revolutionary upsurge as a political *dégringolade:* a spontaneous disintegration of political authority. Its most striking manifestation was the collapse of that system of regional power-holding which had dominated Chinese politics generally, and the Chinese salt administration with them, since the T'ung-chih restoration. Considered at its simplest, the revolution consisted of successful attacks by mutinous soldiers, or other groups of revolutionaries, on the yamens of the governors-general. The ultimate target might be imperial, but the immediate objective would be viceregal, and the governors-general were powerless to avert the risings. The inability of the yamens to resist attack was, however, only the last episode in a decline of influence which had started well before the revolution and, indeed, had made its outbreak possible. This lessening ability on the part of the greater governors-general to dominate the national political scene in the last years of the empire resulted basically from their being subjected to simultaneous attacks on two levels: from the court and from the provinces. It will later be shown that salt affairs were present in the conflict at both levels.

Since the Sino-Japanese war, the imperial government in Peking had been pursuing a deliberate policy of eroding the power of the governors-general by military and fiscal centralization and by frequent changes of personnel. The Mukden correspondent of the *North China Herald* observed that "One of China's greatest mistakes seems to be the constant removal of able officials, especially those who are in high position." [1] As a result, the kind of personal ascendancy in a region, which had been built up over long years by such figures as Li Hung-chang, Chang Chih-tung, and Ting Pao-chen, was now impossible. In October 1911, none of the three Yangtze governors-general — Liang-kiang, Hu-kuang, Szechwan — had held office for as long as three years, and one, Chao Erh-feng of Szechwan, took up his appointment only in the summer of that year. Chao Erh-hsün in Manchuria and Chang Ming-ch'i in Canton were also new to their jobs in 1911, and so were Yü Ch'eng-ko of Hunan and Yang Wen-ting of Shensi among the governors: all of these were key areas in the developing crisis, in which discontinuity of policy and uncertainty about their tenure, on the part of subordinate officials, had serious consequences.

Besides this attack from above, the governors-general had to face a more damaging one from below. A marked feature of politics in the prerevolutionary decade, especially in Kiangsu, Chekiang, Hunan, and Szechwan, was the progressive alienation from the yamens of the local leadership. It was this defection which made the governors-general so vulnerable to the military *pronunciamentos* of the revolution when it came.

Even before the outbreak of the revolution on October 10, however, opposition to the imperial authorities was being voiced through the new provincial assemblies; its leadership came predominantly from officials and degree holders, the very people who, in the days of Tseng Kuo-fan and Li Hung-chang, had been the basis of regional power. Thus, in Szechwan, of ten leaders of the railway protection league arrested by Chao Erh-feng, five, including the president of the provincial assembly, held posts in the ministries in Peking and three others were *chü-jen;* in Hunan, the leadership of the railway agitation was described by the *North China Herald* as belonging to "the very cream of Hunan social life. Everybody who is anybody was either present or was represented." The president of the provincial assembly, T'an Yen-k'ai, was the son of a former Canton governor-general, a 1904 *chin-shih,* and Hanlin compiler; Chang Chien, too, optimus (*chuang-yüan*) in 1894, belonged to this category.[2]

The salt monopoly was deeply involved in the friction between the court and the governors-general. Duke Tsai-tse's creation of the department of salt administration (*yen-cheng yüan*) in October 1911, even if, as was argued in the last chapter, not immediately effective, served notice on the regional power-holders that they were going to lose control of a crucial source of revenue. Under the 1898 Anglo-German loan agreement, 1,800,000 taels of salt revenue were transferred from the local authorities to the Maritime Customs, a body whose primary loyalty was to Peking, while the currency reorganization and Manchurian industrial development loan agreement of April 1911 planned to siphon off a further 2,500,000 taels for a loan which would benefit only the central government.

However, in a disintegrating political context, action for the most part took place on the local rather than the national scene. The issue over which the local leadership most commonly came into conflict with the regional power-holders focused on who should control the new railways, in Chekiang, for example, and in Kwangtung and Szechwan. The importance of the salt revenue in the public finance of the late empire, particularly in connection with railway building, ensured that here too was a significant, if secondary, source of friction. The provincial and national assemblies were beginning to claim the right to a say in the raising of taxes on salt, in Honan specifically in connection with the railway. Here, the provincial assembly passed a resolution in favor of a contribution for railway building to be levied partly on the salt merchants and partly on the local districts. The national assembly also made its views known on the regulation of salt taxation, for example in connection with salt prices charged in Yunnan and north Shansi. It was, however, in the lower Yangtze and in Szechwan that the role of the salt in breaking down the former solidarity between local interests and high officials was most in evidence.

In eastern Chekiang, in the regions of Ningpo, T'ai-chou, and Wenchow, fisheries were a major industry, supplying much of the protein in the diet of the Yangtze delta region. In 1921, the customs commissioner described Ningpo as one of the world's biggest fishing centers, and already before the revolution, the Chekiang fishing industry employed 200,000 people, catching annually half a million piculs of fish. The fishing industry was a large consumer of salt, every picul of catch requiring 2 or 3 piculs of salt depending on the variety of fish, but it was badly served by the traditional salt system with its monopolies and inelastic sales. Since their catch was unpredictable, the fishermen needed to carry a margin of reserve salt, but this the monopoly wholesalers were unwilling to supply, being frightened that the fishermen would dump any unused salt on the inland markets, where the salt merchants kept the price up only by limiting the import. This fear may have been justified, but the legitimate interests of the Chekiang fishery industry were, nevertheless, in conflict with the restrictionist character of the traditional salt gabelle, as analyzed in Chapter 1.[3]

In 1903, these interests were taken up by Ching Pen-po, a local Chekiang official. The fisheries themselves organized a public company, a sort of producer-consumer cooperative, which proposed to acquire the local monopoly of salt sales and buy up increased quantities of salt for the fishermen's use. They offered the government ten times the amount of tax guaranteed by the previous monopolist, but this offer was refused by the governor of Chekiang and the Hangchow salt controller, under pressure from the salt merchants. The merchants were frightened that increased sales in one area would upset the regional balance of the Liang-che monopoly.

Ching Pen-po then joined forces with Chang Chien in Kiangsu, who had been commissioned by the Liang-kiang governor-general Wei Kuang-tao, in his capacity as superintendent of trade for the southern ports (*nan-yang ta-ch'en*), to organize a fishing company of the seven coastal provinces generally in order to meet Italian competition in the fishing business. A proposal similar to that made to the Chekiang authorities was now submitted to the Liang-kiang governor-general, only to be turned down by the same alliance of the salt merchants acting through the salt controller. Chang Chien at once widened the area of controversy by proposing to the board of revenue a comprehensive plan for the reform of the salt administration, which the board referred to Chang Chih-tung, then at Wuchang. Chang Chih-tung pronounced unfavorably on the scheme, as a result of advice tendered by K'o Feng-shih, superintendent of the opium excise and himself a former Liang-huai salt controller.

Salt affairs had thus led Ching Pen-po to run afoul of three sets of regional authorities, a conspicuous case, therefore, of the alienation of a local official from the duly established authorities of the region. He could comfort himself only with Chang Chien's prophetic remark that within ten years the government itself would have changed and reform could continue.[4]

In Szechwan too, the gabelle was a factor in the deteriorating relations between the governor-general and provincial opinion. Salt taxation was rising in the decade before the revolution: receipts on salt destined for Yunnan, Kweichow, and adjoining Szechwan districts rose from 2,288,944 taels in 1900 to 3,739,663 taels in 1909, while the Szechwan levy on Tzu-liu-ching salt exported to Hupei increased from 6 cash a catty in 1901 to 11 in 1911. These increases injured the salt trade in which many Szechwanese were interested, but more harmful from the point of view of the consumer was the additional tax imposed by Chao Erh-hsün in 1908 to make good the revenue lost through the suppression of the opium trade. The correspondents of the *North China Herald* in Chengtu in September 1911 believed that the tax increases were of particular importance in inducing the lower classes to support the railway agitation, which had been begun mainly by the gentry. In November of that year, before the revolution had yet been successful in Szechwan, the British consul reported that the railway protection league was promising a reduction in the price of salt.[5]

The man who linked the railway and salt agitations in Szechwan was Teng Hsiao-k'o. Ching Pen-po described him as the first to begin salt reform during the revolution, and included his letter of resignation from the Szechwan military government in 1912 in his own history of the salt reform movement.[6] In this letter Teng refers to a pamphlet he had written before the revolution, criticizing Chao Erh-hsün's proposal to extend government transportation of salt to the whole of Szechwan. European sources on the other hand frequently link Teng to the railway agitation of the time. A third secretary in the ministry of finance in Peking, Teng was described on his arrest along with other railway agitation leaders as "a most outspoken agitator,"[7] and the British consul in Chungking believed that the tough governor-general Chao Erh-feng wanted him out of the way, a view shared by a missionary correspondent of the *North China Herald.* With the success of the revolution, both of Teng's goals were attained: railway nationalization in Szechwan was abandoned, and salt taxation in the province was reduced by half.[8]

The revolution itself, in the sense of the overthrow, town by town, province by province, of the constituted authorities of the empire, was the work of micropolitical bodies: the railway protection league in Szechwan, the chamber of commerce at Shasi, mutinous soldiers at Changsha, the "dare to die" terrorists at

Shanghai, the Ko-lao hui at Sian. Compared to the initiative showed by these groups, the macropolitical bodies on the scene — the court, the republican government, Yüan Shih-k'ai's ad hoc administration — simply responded to events without controlling them.

After the abdication, the basic problem was whether these small-scale movements could be reintegrated into an effective national government: failing that, their original centrifugal momentum might continue the process of disintegration. If the salt issue contributed to the fragmentation of Chinese politics, which preceded the revolution, it also contributed to the repeated failure of attempts to resynthesize the resulting elements in order to form a stable basis for the new republic.

This occurred first at the local level. Thus Teng Hsiao-k'o's resignation from the Szechwan board of salt (*yen-cheng pu*) late in 1912, over differences of future policy, divided the republicans of that province, while the resignations of Chuang Sung-fu, Fan Kao-p'ing, and Ching Pen-po from the Chekiang salt bureau (*yen-cheng chü*), because the military governor Chiang Tsun-kuei had refused to accept their policies, had the same effect in Chekiang. However, it was in national politics that salt affairs had their most significant impact, salt reform playing a central role in the development of Yüan Shih-k'ai's relations with the two most important political parties: the Kuomintang and the Kunghotang.

That republican politics were early focused on the salt question was due to the reorganization loan negotiations. The traditional fiscal system had collapsed during the revolution, and it was essential for Yüan Shih-k'ai, if he was to have any chance of consolidating his position, to have a foreign loan on a massive scale. Since the Maritime Customs was already fully mortgaged and the future of the land tax and likin was uncertain, the salt revenues were in effect the only possible long-term security for so large a loan, but it was doubtful that the foreign bankers would accept them in their unreformed condition. Salt reform was therefore in the air, and Chang Chien, the doyen of the reformers, accepted an invitation from Yüan Shih-k'ai to go to Peking in the autumn of 1912 to devise a national plan. Before departing from Shanghai, he hammered out a common program with Ching Pen-po and the Chekiang group of reformers, which was translated into English and published under the title *A Plan for the Reform of the National Salt Administration.* The details of this reform program will be discussed in Chapter 6; here the political consequences for Yüan's relations with the Kunghotang and Kuomintang require notice.

Once in Peking, Chang Chien's plan immediately ran into opposition from Chou Hsüeh-hsi, minister of finance, who refused to adopt it as official policy. In salt matters at least, Chou was a conservative, "much prepossessed in favour of the former system of administration," as Dane put it, while Ching Pen-po believed that he was motivated solely by the fact that he himself had 400,000 taels invested in Liang-huai *yin* rights.[9] In view of the fact that Chou was a prominent member of Yüan's government, his hostile reception of the salt reform plan angered Chang Chien and his sympathizers, and weakened their confidence in the president's intentions.

The differences between Chang and Chou were not only personal, nor confined to salt affairs. They represented different political groups, offering alternative bases of support to Yüan Shih-k'ai. Chou Hsüeh-hsi, former Ch'ang-lu salt controller, manager of the Lanchow coal mining company, and related to Yüan by marriage, belonged to the president's personal entourage, the circle he had collected as Chihli governor-general. Chang Chien on the other hand, president of the Kiangsu provincial assembly and close friend of men such as T'ang Shou-ch'ien and Ch'eng Te-ch'üan, president of the Chekiang assembly and military governor of Kiangsu respectively, represented that class of provincial notables whose defection had brought down the old governors-general. It was these men who in 1912 formed the nucleus of the Kunghotang. Yüan's ultimate inability to win their support was an important factor in his overall failure, and salt reform was one of the issues which precipitated it.

The salt issue also affected Yüan's relations with the Kuomintang or, as it was named before August 1912, the T'ung-meng hui. In order to defeat Chang Chien's proposals for reform in the provisional senate (*lin-shih ts'an-i yüan*), Chou Hsüeh-hsi needed Kuomintang support, and he accordingly entered into an alliance with this party. Ching Pen-po, who had nothing good to say about Chou, regarded this as a purely opportunist maneuver, but it may be conjectured that, following the failure to ally with the Kunghotang, Chou's move was part of an alternative plan to infiltrate and capture the Kuomintang. This desire for a coalition was shared by certain of the Kuomintang leaders, Huang Hsing in particular, who in September 1912 advocated that Yüan become the party leader. Ching Pen-po has said that the Chao Ping-chün cabinet of 1912 was a "quasi-Kuomintang cabinet." [10]

Here, too, salt played a divisive role. Chou Hsüeh-hsi, now a party member, tried to use the alliance to get his own conservative policy for the gabelle endorsed by the Kuomintang. In this attempt he was thwarted by Sung Chiao-jen, who warned that to support the salt merchants would be political suicide. Sung thus destroyed a possible basis for the whole entente. This obstruction must have been one more count against Sung in the complex of factors which led to his assassination in March 1913 and to the definitive break between Yüan and the Kuomintang.

The course of political history from 1911 to 1913 may, therefore, be viewed as an ongoing process of distintegration, broken only by Yüan Shih-k'ai's attempt to resynthesize the fragments of the shattered power structure. In this process, the salt monopoly played a significant part as one of several sources of friction and eventual fragmentation.

The gabelle's own history in the revolution reproduces this general pattern of disintegration and tentative recovery. The first severe challenge to the existing order was provided by the intensification of disorder and violence throughout 1912 and 1913.

Disturbance of the peace in the sense of riot, armed smuggling, and minor insurrection was endemic in the prerevolutionary salt administration, just as disorder in general was to be found in all the institutions of traditional China, to an extent unfamiliar in modern societies. The salt laws themselves, their complexity giving many opportunities for law-breaking, positively fostered disorder: they were, as Teng Hsiao-k'o put it, "lawless laws." [11] The *shih-lu* for the prerevolutionary decade abound in references to the "owls," as smugglers were known. Early in 1900 the governor of Anhwei memorialized that the "owls" were causing trouble in the Ho-chou area and requested regular troops for use against them; in 1902 an anonymous memorialist stated that in Kiang-nan the salt owls in their small craft were "numerous as a forest of masts" and were robbing people and holding them for ransom; while in 1908 Liang-kiang governor-general Tuan-fang reported that bands of salt smugglers up to 300 strong in the Chekiang–Kiangsu border region were indiscriminately attacking likin barriers, churches, and schools.[12]

Prevalent as disorder was, it can scarcely be doubted that the revolution of 1911 multiplied both its degree and its incidence, producing a total breakdown in the functioning of the traditional system. Increased disorder can be documented from many areas. In Szechwan, Teng Hsiao-k'o reported that during the winter of 1911–12 the salt export from the four main centers of Chien-wei, Lo-shan, Fu-shun, and Jung-hsien was interrupted by troops from Yunnan. Similarly, the *North China Herald* correspondents in Szechwan made frequent reference to violence affecting the gabelle during the revolution: its Chengtu correspondent, on September 7, 1911, stated that "robbers" were attacking salt offices, likin and tax stations, while on December 9 the salt tax offices at P'eng-hsien were reported wrecked by *banditti* organized by the Ko-lao hui. At Yangchow preparations for the defense of the salt commissioner's house were reported, while at Ichang revolutionaries seized the person of the salt taotai. In Anhwei, the customs commissioner in charge of the salt likin collectorate at Ta-t'ung described a battle on December 8 between rival groups of revolutionaries, which took place on the grounds of the commissioner's house, for the control of the local salt revenues. Yang-chow, Ichang, and Ta-t'ung were all major centers for the collection of the salt tax.[13]

Under these strains, the traditional salt administration ceased to operate effectively. Dane, discussing the revenue position for 1912, declared that "The administration was disorganised by the Revolution of 1911, and the Provincial Governments ceased almost entirely to remit revenue to Peking," [14] a view shared by a government spokesman in a speech in June of that year to the provisional senate. In Szechwan in the same year, Teng Hsiao-k'o argued against any attempt to restore the ancien régime by pointing out that "wholesale capital has been dissipated, routes are obstructed, tax offices have been destroyed by fire, and the merchants have fled." [15] Dane, in his account of Huai-nan, referred to "the disorganisation of the Huai Yinti system of administration, which was caused by the Revolution," stating that "during the trouble of 1911 to 1913 the Huai Yinti system very nearly broke down." [16] In Huai-pei too, he found that "The administration appears to have broken down under the stress of the Revolution in 1911," [17] and in 1913 only 67,870 *yin* were being transported instead of the pre-1911 figure of 360,000.

A temporary breakdown, such as occurred between 1911 and 1913, was not in itself untraditional. It could be viewed as part of a dynastic cycle, and the perennial recuperative powers of the tradition could be relied upon to restore the system in due time. What was untraditional was the amplification of this breakdown by two additional factors, coincident with it: the rise of a Chinese salt reform movement which consciously rejected the traditional order and which implemented a number of changes that further weakened the old system; and the imposition of foreign control by the reorganization loan. These two factors effectively interrupted the cycle and prevented the revival of the old regime.

The Salt Reform Movement, 1911–1913

The disorder accompanying the revolution had produced a functional breakdown in the salt administration. The new salt reform movement, diverging from traditional patterns of thought, carried the process a stage further and damaged the nerve center of the old regime. Though the constructive achievements of the new movement were limited, by changing the intellectual and institutional setting, it made any restoration of the traditional system difficult, if not impossible.

The origins of the salt reform movement. In common with most political groups involved in the revolution of 1911, the salt reform movement was at first disconnected and localized. There was no single grouping, only a number of regional centers, reflecting different backgrounds and aiming at different objectives. Six major foci of reform coexisted: two in Chekiang and one each in Kiangsu, Szechwan, Fukien, and Manchuria. There are also indications of a salt reform movement in Kwangtung, while arguments in a Chekiang manifesto imply the existence of a third school there, which wanted to abolish the salt tax altogether. These movements approached salt reform from differing standpoints and varied in their relationships with the prerevolutionary provincial authorities. Three cases in particular illustrate this.

In Kiangsu,[18] the interest shown by Chang Chien in salt was the unforeseen by-product of another undertaking. He and his group of reformers were first involved in the problems of the gabelle as far back as 1900, as a result of their plans to convert land previously used for salt production to cotton cultivation. These were opposed by the Liang-huai salt controller and the powerful salt-producing interests, the yard merchants (*ch'ang-shang*). Probably they were worried that competition for land would raise the price of reeds for fuel and salinity intensification, and hence worsen their already weak competitive position vis-à-vis Huai-pei solar-evaporated salt. Chang Chien's quarrel then was more with the officials and producers than with the distributors, the transport merchants (*yün-shang*). Its origin was competition to utilize the same means of production, rather than a consumer protest against a monopoly of distribution.

In Chekiang[19] on the other hand, it was the interest of a particular group of consumers, the fishing industry, which led Ching Pen-po into the salt reform movement in 1903. The east Chekiang fisheries could not obtain adequate supplies of salt, and Ching Pen-po supported their proposals to acquire the monopoly for themselves. His particular enemy was the merchant who held the distributing monopoly in the Ningpo region. This antagonism reflected the fact that in the Liang-che salt division, indeed in south China generally, official control was weak and that it was the distributing merchants who were the dominant element. Ningpo was a conspicuous example of this situation. In 1913, the foreign inspector described the position as follows: "The salt business in these districts was entirely managed by the said merchant who kept a special preventive force against smuggling, established his own controlling offices within these districts, and had private arrangements with the salt works where he obtained salt." In addition, while Liang-huai was "the most highly organised" salt division in China, Liang-che was "exceedingly complicated and . . . very inefficient."[20]

In Manchuria,[21] the context was different again. There, as the author of one reform program pointed out, there were no entrenched merchant monopolies. In Kirin and Heilungkiang, the authorities had in 1908 adopted the now fashionable policy of "official transport" (*kuan-yün*) and planned to extend it to Feng-tien. From the reformers' point of view therefore, it was only a question of seeing that this policy was actually carried out. Unlike the reformers in Kiangsu and Chekiang, those in Manchuria were concerned not with sectional interests but with overall administrative efficiency and the welfare of the community. As the current political cliché put it, "to remove corruption and hindrances is to benefit the people; to increase sources of taxation is to enrich the state." In this aim, the reformers felt they were working with, rather than against, the officials. The apparent absence of antagonism reflected perhaps the solidarity of a frontier community which had had a run of good governors-general (Hsü Shih-ch'ang, Hsi-liang, Chao Erh-hsün) and where the revolution itself had taken the form of an orderly transference of power without the violence found elsewhere.

The different movements also advocated differing remedies for the defects of the ancien régime. In Kiangsu, Chang Chien supported the policy known as "taxation at source and minimal interference thereafter" (*chiu-ch'ang cheng-shui, jen ch'i so chih*).[22] This policy limited the role of officials to tax collection at source, but did not exclude the continued existence of private monopolies of distribution. In Szechwan, Teng Hsiao-k'o adopted the more radical position of taxation at source and free trade thereafter (*tzu-yu mao-i*).[23] In Chekiang on the other hand, Ching Pen-po stood for the opposite policy of official control down to the level of retail (*chiu-ch'ang chuan-mai*), which was also adopted by the reformers in Manchuria and Fukien.[24] Chekiang opinion, however, was not unanimous in its support of this policy, for after October 1911 the salt bureau set up by the military government under Chuang Sung-fu and Fan Kao-p'ing, with Ching Pen-po as secretary, adopted the weaker policy of official purchase from the producer plus subsequent distribution by monopoly merchants (*kuan-shou shang-mai*), a compromise which would have retained much of the traditional structure. The salt reform movement therefore displayed a broad spectrum of origin and aim.

The intellectual development of the salt reform movement. The salt reformers' most positive achievement in this period was to break out of the confining ideology of the tradition at certain critical points. This judgment is not vitiated by the fact that the reformers saw themselves for the most part as arguing along conventional lines. The reality was that the analytical distinctions they brought to the discussion were deadly for traditional styles of reasoning. To substantiate this, we need to examine their ideas of their own ideological stance and also the crucial moments where, without fully realizing the significance of their arguments, they introduced new ways of looking at old problems.

The starting point of their analysis was a common hostility to the *yin* system (*yin-chih*). In criticizing it, the reformers assumed themselves to be part of a continuing discussion. This is shown particularly in the details of their argument, their citation of earlier authors, and their view of their own intellectual ancestry.

What did the reformers mean by the *yin* system? The term *yin* had more than one meaning: it could refer to the certificate or title deed which permitted a merchant to participate in the salt trade, or, alternatively, to the quantity of salt covered by such a certificate, that is, a unit or measure of salt, translatable into piculs and catties, and varying in different parts of the country. However, it was in neither of these senses that the reformers used the term. They used it to refer to the system whereby a *yin* or certificate (1) assigned to the holder a sales monopoly within a specified area, (2) designated the producing region from which he must obtain his supply, and (3) defined the quantity of salt covered by one utilization of the certificate, which could, however, be used again, provided the original consignment had been sold.

The reformers did not object to the monopoly as such, but to the smallness of the units involved. Ching Pen-po wrote: "If the salt merchants of the whole country were to form themselves into a single monopoly company like a trust, then the general public would only have to put up with financial hardship and they would be free of the above-mentioned *yin* subdivisions (*yin-chieh*)." [25] Similarly, the manifesto of the Chekiang salt bureau stated that "the former *yin* divisions (*yin-ti*) took the hsien as a unit. This area is too small," [26] and it proposed to redivide the province into seven larger monopoly regions. The small units, it was felt, aggravated the inelasticity inherent in the *yin* system. *Yin,* it was true, could be used more than once in a year,[27] so that there was no ceiling on sales of salt. As the Chekiang fishermen found, however, the monopolist could not be compelled to increase supplies if he felt it against his interests. The smaller the unit, the less latitude there was for the individual consumer.

This extreme subdivision of monopoly rights has not been sufficiently emphasized, because it was not fully developed in the best-known salt division, Huai-nan. There, while the *p'iao,* or entitlement to trade between Yangchow and the up-river ports, specified the quantity to be shipped and whether it was to go to Anhwei, Kiangsi, Hupei, or Hunan, it did not specify the saline from which salt was to be obtained or a precise destination within the up-river province. Thus, within the limitations of the *p'iao,* trade was relatively free and quantities sold in the different sales areas fluctuated. In Anhwei (the so-called Wan-an lit., "Wan ports"), "The final destination of the salt, *i.e.* the place of sale in the *an* district, is determined by the Tu Hsiao head office at Tatung according to the existing demand at the different places." [28]

Huai-nan, however, was exceptional: in other areas, the scale of the monopoly was smaller, each district (hsien) constituting a distinct monopoly area with a fixed number of *yin.* In Shantung, "For each of the Yin and Piao districts a fixed number of Yin and Piao licenses were issued. The number of merchants for each consumption district was also fixed, and the salt centres from which salt might be packed were likewise fixed for each merchant." Or, again, "In Shansi the trade in salt was restricted. Each district was allocated to a particular merchant . . . who maintained his own shops for retail sale and alone was permitted to sell salt within the district." Districts (hsien) could even be subdivided into separate monopoly areas. Teng Hsiao-k'o wrote, "people may belong to the same hsien yet belong to different monopoly merchants." For example, in Kwangtung, Fatshan (Namhoi hsien) was divided into six monopoly areas and Canton city (P'an-yü hsien) into eight.[29]

This system was the primary target of the reformers. It is clear, however, from their writings that they did not regard their criticisms as anything new. On the contrary, they regarded themselves as dealing with a problem which had divided salt reformers into two camps since late Ming times.

On the one hand, they distinguished a school of thought descending from Li Wen (secretary of the Grand Secretariat, 1644–

1646) and Ku Yen-wu (1613–1682), which opposed the *yin* system and advocated the policy of taxation at source and minimal interference thereafter. Its arguments they summed up in a cryptic remark which the Manchurian reform program attributed to Ku Yen-wu: "If all salt is smuggled, then no salt is smuggled" (*t'ien-hsia chieh ssu-yen, t'ien-hsia chieh kuan-yen*).[30]

This may be interpreted as follows: a defect of the *yin* system, already present in Ku's day and still being attacked by Ching Pen-po and Teng Hsiao-k'o, was that it created an artificial category of crimes, since it laid down that salt assigned to hsien A could not legally be sold in hsien B. Furthermore, there was every incentive to commit these crimes. Under the old system, salt paid tax twice, at the source and again in the district of consumption. Salt destined for hsien A and smuggled instead into hsien B paid only one installment and could thus compete successfully in hsien B against hsien B salt which had paid the two required installments. In this way, the revenue was defrauded and the monopoly system broke down. Ku's remedy was to do away with the *yin* divisions and with any further taxation once the salt left the field. Once it had paid tax at the field, any salt would be free to compete in any market. It would be unauthorized or "smuggled" in terms of the old system, yet all of it would now be taxed and legitimate. The clandestine trade would be eliminated and revenue would increase.

On the other hand, as the reformers saw it, there was a school of thought associated with T'ao Chu and Feng Kuei-fen, the leading nineteenth-century theorists of reform in the salt administration. Its arguments, too, they summed up in an aphorism which the Chekiang reform program attributed to Feng: "If all salt is legitimate, then no salt will be legitimate" (*t'ien-hsia chieh kuan-yen, t'ien-hsia chieh ssu-yen*).[31]

This was intended as a rebuttal of Ku Yen-wu, with the counter-argument running like this: under his system, all salt once it had paid tax at source would be legitimate, wherever it was found, and there would be no means of distinguishing between legal and smuggled salt in the hands of the consumer. Since it was not in fact possible, Feng argued, to control and tax all salt at source for a variety of reasons, taxation at source and minimal interference thereafter would have the effect of encouraging the illegal trade at the expense of the legal. Ku's system was designed to universalize legitimate tax-paid salt; Feng Kuei-fen argued that it would only universalize smuggled salt. The clandestine trade would increase and revenue would decline.

Towards this traditional controversy, the reformers did not, it is true, all take the same attitude. Chang Chien, for example, accepted Ku Yen-wu's program wholeheartedly, while Ching Pen-po and the Chekiang group accepted his diagnosis but not his cure. The Chekiang manifesto reproduced once again an old argument of T'ao Chu's: if taxation was to be paid at source, who was to pay it? The producers could not, for lack of resources, being mere artisans, and if the merchants paid, this would involve no real change of system.

From the prominence given to the question of the *yin* system, it may be concluded that the salt reformers of the revolutionary period saw themselves as part of a long tradition of debate: Liu Yen of the T'ang and Ts'ai Ching of the Sung still symbolized to them live issues. They did not, however, remain confined within this tradition and at three crucial points achieved a breakthrough.

In the first place, the Chekiang manifesto felt it necessary to justify the existence of a salt tax at all, and this by reference to three general fiscal principles: taxation should, it argued, be (1) equitable and universal; (2) low in incidence per person but high in overall receipts; and (3) easy to administer. These three criteria are reminiscent of Adam Smith's four principles of taxation in the *Wealth of Nations,* which had been available in Chinese in Yen Fu's translation since 1900.[32] Whether the Chekiang reformers were already influenced by Smith or not, their manifesto shows both a new awareness of public finance as an autonomous discipline, distinct from general political science, and an acknowledgment of the fact that salt problems are not *sui generis* but must be judged according to general fiscal principles. The frame of discussion had been altered by particularizing the general principles and departicularizing the salt problems.

Even when the reformers did appeal to general political values, they felt that these had changed. Teng Hsiao-k'o argued that "the salt laws' official transportation and monopoly were suitable for a monarchy; but the excellent law of taxation at source accorded with the republican trend." Ching Pen-po was more precise: the *yin* system with its subdivision was, he held, a kind of feudalism; 2,000 years after the ending of political feudalism, the salt merchants' fiefs still existed and ought to be destroyed along with the monarchy.[33]

Finally, the reformers gave a new significance to the conventional four-word phrases in which the traditional gabelle expressed itself. In the language of the older system, terms such as "official supervision, merchant sales" (*kuan-tu shang-hsiao*), "official transport, official sales" (*kuan-yün kuan-hsiao*), and "taxation at source and minimal interference thereafter" (*chiu-ch'ang cheng-shui, jen ch'i so chih*) appear at first sight to refer respectively to systems operating through licensed merchants, to complete state monopolies, and lastly to free trade in a Western sense. They seem, that is, to fit into the familiar dichotomy of *étatisme* and laisser-faire.

In fact, however, the meanings of the Chinese terms were much more limited: they were describing administrative procedures rather than different economic systems, and the differences between the procedures were insignificant in terms of economic theory, as has been noted in Chapter 1. Furthermore, their function was as much normative as descriptive, designed to win approval for a course of action by ascribing it to accepted and traditional modes of procedure. With the reformers, however, the traditional phrases began to acquire a more "Western" meaning. Under the taxation-at-source slogan, Teng Hsiao-k'o was introducing genuine free trade, while in Manchuria the reformers saw the traditional official transportation system as a first step towards a fully nationalized industry.

In brief, traditional thinking about salt problems had a particular logical structure. It consisted on the one hand of general propositions, moral or metaphysical in character, such as "the state's taxes and the people's livelihood" (*kuo-chi min-sheng*), and on the other of almost esoteric formulae with a limited reference to the administrative procedures of the gabelle. What it lacked was language operating in the middle range between the classics and the handbooks of the clerks, the sphere precisely of modern politics, economics, and public finance. It is because there was a growing awareness in the writings of the reformers of this middle range that they can be regarded as no longer wholly traditional but as modernizing. At the level of thought therefore as well as of function, the old regime was breaking down in the period 1911–1913.

Organizational development of the salt reform movement. The same modernizing tendency can be seen in the organization the reformers established at the end of 1912, the Salt Administration Discussion Society (Yen-cheng t'ao-lun hui) or Salt Society, as it may be abbreviated.

Until December 1912, the salt reform movement was decentralized, lacking a common program or national organization. Even within a single province, reformers had little contact with each other: thus Ching Pen-po had not met Chuang Sung-fu when they both joined the Chekiang salt bureau in 1912. The first task of the reformers, as of republican China as a whole, was to combine the disparate elements of 1911 into a united whole. In this, thanks to the organizing ability of Ching Pen-po, the reformers had a measure of success: the formation of the Salt Society demonstrated in a specialized field what might have been achieved in national politics more generally.

The immediate occasion for the formation of the society was the discouraging situation of Chang Chien and Ching Pen-po in the autumn of 1912. Although summoned to Peking by Yüan Shih-k'ai himself, they found themselves opposed by Chou Hsüeh-hsi who, through his alliance with the Kuomintang, commanded a majority in the provisional senate. In addition, they realized they had against them the vested interests of the traditional salt officials and merchants, a total of several million people as they reckoned it. Their hopes were set on the meeting of parliament the following year, and it was with the aim of influencing its members and public opinion generally that Ching Pen-po proposed to establish a society which would publish a magazine devoted to salt affairs. The society did not adopt a definite program, but the first number of its magazine published Chang Chien and Ching Pen-po's own plan, and it was assumed the contributors to the magazine shared a common dislike of the status quo, the *yin* system in particular.

What kind of a body was the Salt Society? Despite an initial membership of several thousands, it was never a mass movement, and though Ching Pen-po was concerned to arouse public interest and took note of newspaper comment on salt affairs, the society was more concerned with official than popular opinion. Ching's editorial narration of events was focused on parliament, and for him Yüan himself and the minister of finance were the chief dramatis personae. Bills were introduced, and support was canvassed for: after the murder of Sung Chiao-jen and the breakdown of the Chou Hsüeh-hsi–Kuomintang alliance, 200 Kuomintang members of parliament entered the society. The Salt Society was an elitist rather than a popular body.

The open formation of a pressure group of this kind nevertheless represented another break with tradition. Under the empire, since the Yung-cheng Emperor's essay on factions (*P'eng-tang lun*) of 1724, open political associations of any kind had been frowned upon, and salt reformers wishing to propagate their views expressed themselves through such genre writings as essays published in the *ching-shih wen-pien* and through regional treatises on the salt laws (*yen-fa chih*). The revolution of 1911 broke down those restrictions: the Salt Society was not a government association, yet it could function openly and campaign freely through its magazine. Salt policy could now be a matter of open debate. So long as Chou Hsüeh-hsi remained finance minister, the society's influence with Yüan Shih-k'ai was limited, and its hopes of obtaining control of the country's salt administration were dashed by the signing of the reorganization loan agreement. But its very existence altered the circumstances of all salt reform: in however limited a fashion, the Salt Society brought the question of the salt administration into the court of public opinion. This was a transformation as striking in its own sphere as the innovations in argument noted earlier.

Salt Reforms, 1911–1913. At the same time that the reformers were changing the climate of salt reform, a number of practical administrative changes were being introduced. These reforms, inspired by the new climate of opinion, further undermined the former structure.

The overall picture of these changes in the provinces is summarized in Table 10. Many of these were introduced by local reformers acting in the first flush of the revolution and preceded the formation of the society in the autumn of 1912. This was the case in Huai-nan, where Chang Chien attempted to apply his taxation-at-source principles when he became Liang-huai salt commissioner; in Szechwan, where Teng Hsiao-k'o proclaimed free

Table 10. Changes in the salt administration, 1911–1913.

Division	Nature of change
Manchuria	No change
Ch'ang-lu	Taxation simplified, official control extended
Shantung	No change
Ho-tung	No change
Huai-pei	Official control extended
Huai-nan	Taxation at source
Liang-che	No change
Fukien	Official control extended
Kwangtung	Free trade
Szechwan	Taxation at source, free trade
Yunnan	No change
Northwest	Official control extended

Source: *YWKMS*, p. 7; *YWKMS*, Appendix, pp. 9–10; Dane Report.

trade and cut tax rates by half; in Kwangtung, where the monopoly of the so-called farm merchants (*kuei-shang*) was ended; and in Fukien, where on the recommendation of Liu Hung-shou, who subsequently became salt commissioner, a complete government monopoly was introduced.[34]

Other reforms, however, were begun at the instance of the ministry of finance, where the reforming impulse imparted by Duke Tsai-tse in the last months of the empire was continued by Yüan Shih-k'ai's ministers. For example, the principal change in Ch'ang-lu, the establishment of the Honan government transportation office "under the direct orders of the Ministry of Finance," in 1913, was simply the continuation of a process begun by the takeover of merchants' rights in this area in 1911. The establishment of the Jehol and K'ou-pei transportation offices which affected Mongolian salt was also in line with the late Ch'ing policy of extending official control in these regions via the Peking–Mukden and Peking–Kalgan railways.[35]

However, the most striking feature about the reforms, taken together, was that they did not embody any one consistent common principle. On the one hand, there were liberalizing reforms as in Szechwan, Huai-nan, and Kwangtung; on the other, there were changes which implied more official control, not less, as in Ch'ang-lu, Huai-pei, Fukien, and the northwest. The move to base salt administration on definite fiscal principles was in itself a sign of a more modern approach, but the diversity of the principles adopted meant that no solid basis for a uniform national system had been laid; indeed, the lack of systematization was now more glaring than ever.

The Reorganization Loan Agreement

Ambiguities of a different kind affected reforms at the national level. Here, early in 1913 Chou Hsüeh-hsi introduced two reforms which, when properly carried out, were to become the basis of Sir Richard Dane's reform. First, an attempt was made to centralize tax receipts and prevent local appropriations of any kind. According to a presidential mandate of January 6, 1913, "all salt revenue shall be set aside and held as a special account, and under no circumstances shall the revenue be drawn upon in the form of either loans or appropriations." Second, a new echelon of the salt administration, the inspectorate, was set up to ensure that the revenues collected were in fact paid into the central account, and it was provided that foreigners should be employed parallel to the Chinese at all levels of the new body. The creation of the Sino-foreign inspectorate, the chief instrument of Dane's reform, thus preceded the signature of the reorganization loan agreement by a few months, and foreign assistant inspectors were in fact appointed to Liang-huai and Fengtien in April 1913.[36]

The significance of these reforms must not be exaggerated. All attempts to revive the salt tax in 1912 and early 1913 had proved unsuccessful: *a fortiori* then, any attempt to centralize the receipts must have failed.

As for the inspectorate, the intention of the Chinese government ought not to be taken at its face value. The reorganization loan had not yet been signed, but China had accepted its principles and final signature was being held up only by disagreements among the powers as to the distribution of the advisership posts to be created under the agreement. Since, therefore, the establishment of the Sino-foreign inspectorate was only a matter of time, the Chinese government may have believed that by taking the initiative in setting up the inspectorate ahead of the agreement, they could circumscribe its authority and render it comparatively innocuous. Ching Pen-po certainly believed that Chou Hsüeh-hsi regarded foreign control simply as a means of thwarting the Chinese reformers and preserving the *yin* system in which he was personally interested. This view received some support from the regulations governing the inspectorate issued in January and March 1913. These provided that "All executive administration concerning the production, manufacture, transportation and sale of salt in the different provinces does not come within the scope of the Chief and District Inspectorates' authority," and, in addition, that communications from the inspectorate to the central salt administration (*yen-wu shu*) were to be in the form of petitions (*ch'eng*), while communications in the opposite direction were to be in the form of orders (*ling*).[37] The inspectorate was clearly intended to be a subordinate body.

In the long run, foreign control was to do most to modernize the gabelle. This was not foreseen in 1913, when Dane wrote that "Much scepticism prevailed among both Chinese and Europeans as to the possibility of the salt revenue administration being effectively reorganised." Looking back, it is plain that the importance of the reorganization loan agreement for the gabelle lay not so much in what it was as in what Dane made of it. Nevertheless, the agreement was his starting point, for, by article 5, China was committed "to take immediate steps for the reorganisation, with the assistance of foreigners, of the system of collection of salt revenues of China." [38]

The unprecedentedly large loan — £60,000,000 were originally envisaged — was more than a simple financial transaction, even though the negotiations were conducted by the Peking managers of the foreign banks. It was in effect the first serious effort by the six major powers of the treaty system, namely Britain, France, Germany, the United States, Russia, and Japan, to respond to the situation created by the fall of the Chinese empire, the activities of the bankers at every stage being closely supervised by their ministers in Peking and by their foreign offices at home. The political purpose of the loan was to enable Yüan Shih-k'ai to establish a strong central government, one which would maintain the unity of China and obviate the necessity for extensive foreign intervention, two desiderata which all the powers may be said to have shared in a greater or lesser degree. With considerable skill, Yüan had kept the 1911 revolution Chinese, carrying through the radical surgery demanded by Chinese opinion without civil war and without foreign intervention. Now he required money to continue, but it was essential that he acquire it on terms which would not outrage nationalist opinion and destroy his consensus of support.

The details of the negotiations leading up to the signature of the agreement at one o'clock in the morning on April 27, 1913, have already been described by P'u Yu-shu in his unpublished dissertation, "The Consortium Reorganisation Loan to China, 1911–1914: An Episode in Pre-war Diplomacy and International Finance" (University of Michigan, 1951).[39] However, the greater availability of the British archives in recent years makes it now possible to be more explicit on several problems: the origins of article 5, which created the foreign gabelle; the decision to entrust the gabelle to an Englishman; and the selection of Sir Richard Dane as that Englishman.

Yüan Shih-k'ai first requested a large loan soon after he took office as imperial premier in November 1911. The powers, wishing to remain neutral, refused him assistance, and it was not until February 1912 that serious negotiations began, in the first instance for immediate advances in anticipation of a larger loan to be concluded later. On February 24, E. G. Hillier, Peking manager of the Hongkong and Shanghai Bank, telegraphed his London office: "We propose finance foregoing by means of sterling Treasury bills secured as second charge on salt revenue and redeemable from re-organisation loan later, same being subject to stipulations that immediate steps will be taken to re-organise salt gabelle with expert foreign assistance." [40] These words reappeared only slightly changed in article 5, so that Hillier, the blind banker of Peking, may in a sense be regarded as the father of the whole idea of a foreign gabelle.

Between February 1912 and April 1913, the proposal for foreign participation in the salt administration encountered strong Chinese opposition. The bankers' original demand for the employment "of foreign officials, to administer revenues pledged as security either under maritime customs or under separate but similar organisation," [41] was rejected by the Chinese in June, and their counterproposals offered only the most limited kind of foreign advice and technical assistance.

Faced with this Chinese resistance, described by Sir Edward Grey as "the evident determination of the Chinese Government to resist, at all costs, any foreign administration of the pledged security," [42] the British negotiators were inclined to yield. Jordan was sceptical of the practicability of foreigners operating the Chinese salt administration at all, a view in which Francis A. Aglen, inspector-general of the Maritime Customs, concurred. Charles Addis, London manager of the Hongkong and Shanghai Bank, doubted the wisdom of pushing Yüan too far, while Sir Eyre Crowe, a senior official looking at the problem from the point of view of Britain's *Weltpolitik,* feared that the new foreign gabelle might be a predominantly German institution and be used to undermine Britain's position in the Far East. By September, thanks to these doubts of his advisers, Grey was ready to accept the Chinese counterproposals.

That the banking consortium stood its ground was due to France: to Poincaré and Jonnart, successive foreign ministers, to Stanislas Simon, Paris manager of the French banking group, and especially to Conty, the French minister in Peking. Jordan told Grey on January 20: "It is quite noticeable in all these discussions that the French wish to obtain through their agents a direct control over Chinese administration." [43] Confronted with vague Chinese promises of advisers, it was Conty who on November 12 insisted on a "conseiller étranger qui travaillerait à l'administration centrale de la gabelle." [44] After China had conceded a foreign associate chief inspector in place of a mere audit officer, it was Conty again who maintained that he must be a national of one of the lender countries, and not a neutral from a smaller power who would lack authority to enforce reform. Conty, it is clear, was the person mainly responsible for driving home Hillier's proposal for a foreign gabelle.

So long as foreign participation was to be confined to audit or advice, Jordan and the British government had not been concerned as to the nationality of the person to head it, though they were not displeased when Yüan Shih-k'ai indicated that he would choose a British customs commissioner, Macoun, for the job. It was only when the auditorate was strengthened to an inspectorate and when Conty insisted that the foreign chief inspector must be a national of one of the lender countries that Jordan played hard for a British appointment. The worsening financial situation in China while the loan remained unconcluded raised once more the specter of foreign intervention. Jordan telegraphed Grey on February 14, 1913, that "advisers of to-day will in all probability become commission of control of the future," and Beilby Alston at the Foreign Office minuted on the file that "we shall find the advisers developing into a Turkish or Egyptian 'Caisse de la Dette.'"[45] If the appointment was to carry with it anything of the very real power wielded by the officials of the Ottoman debt administration, Britain could not lightly see it go to any of her rivals.

The crucial moment of the negotiations occurred at a meeting of the ministers of the six consortium powers in Peking on February 4.[46] The issue was complicated by the fact that, besides the salt appointment, the powers had also to allocate among themselves the audit department supervising the expenditure of the loan, and a loan department concerned with future borrowing. These were regarded at the time as of equal importance to the salt inspectorate; as it turned out, however, they were too vaguely conceived to develop effectively.

In Jordan's account, the meeting opened with the Russian minister demanding the salt post for his country, on the basis that Russia was holding the largest share of the Boxer indemnity (a prior charge on the salt revenue), but offering to share it with Britain on a north-south division of authority. Jordan at once sidetracked the proposal by arguing that the salt administration did not lend itself to this kind of division, and that the audit department could be shared much more easily. Conty then offered to share the audit department with Russia, France's closest ally, thus leaving the way open for the appointment of an English inspector. Once the entente powers had settled their differences, the German minister was left isolated, since Japan would support Britain, and William J. Calhoun, the United States minister, was at best neutral, so that the German representative Haxthausen had to aquiesce in Britain taking the inspectorate, contenting himself with the loan department, the least important of the three appointments at issue.

The German government, however, was not so complaisant: it refused to ratify the ministers' agreement and instructed Haxthausen to negotiate again for the chief salt post directly with Jordan. Jordan resisted, though he offered the Germans as consolation a deputy chief inspectorate based in Shanghai, subordinate to the British chief inspector at Peking. In the end, the German government, like its minister at Peking, found itself too isolated to press its claims and accepted Jordan's compromise. Jordan therefore proclaimed the outcome of the talks a victory for England: "I shall never cease to look back upon the appointment of a Britisher to the Salt as one of the achievements of which we have a right to be proud." [47]

With the question of who the Britisher was to be, surprisingly, neither Jordan nor his superiors in London were much concerned. Dane's name was first introduced by the Chinese minister in London, Liu Yuk-lin, who told Alston at the beginning of February 1913 that he was recommending Sir Richard Dane to his government for appointment as adviser on currency reform. This was another Sino-foreign project, which had originated before the revolution and was still being canvassed in 1913, though in the end nothing came of it. Alston then drafted a letter to Jordan at Peking, informing him of Liu's recommendation and asking if Dane would not also be suitable for the salt appointment. Jordan did not reply to this query, probably because he was still deep in negotiations as to whether the salt appointee was to be British at all, but later comments showed him warmly approving.

Indeed, the Chinese government probably invited Dane because it knew Jordan approved of him. On March 4, Yüan formally rejected the demand of the foreign legations that they nominate their nationals to the salt, audit, and loan posts in the way they had agreed at the ministerial conference of February 4. This, however, was only a maneuver to placate Chinese public opinion. Privately, Yüan informed the bankers that he was willing to accept the legation plan if the interest on the loan was reduced from 5.5 to 5 percent, and, receiving encouragement from them, he began appointing on his own authority the men whom he knew the powers concerned would have chosen themselves: N. A. Konovaloff, a Russian, to the audit department, and C. Rump, a German, to that for loans. On April 10, the India Office communicated to the Foreign Office a telegram from Dane at Lahore, saying that he had been offered the salt appointment by the Chinese minister in London and asking for British government approval. This was quickly given: on April 15, the viceroy telegraphed the India Office that Dane would accept, and, on April 18, Jordan reported that Chou Hsüeh-hsi, the Chinese minister of finance, had told him that Dane had accepted.

The only remaining puzzle in Dane's appointment is how Liu Yuk-lin came to think of him in connection with the currency post in the first place. A likely solution is that Liu heard of Dane through Sir David Barbour, financial member of the council of the governor-general of India, 1887–1893, a former colleague of Dane's, and a fellow Anglo-Irishman. It was he to whom Liu later applied for Dane's address. Barbour, furthermore, had in 1911 sat on an international committee of experts in London advising on Chinese plans for currency reform, and one may guess that early in 1913 he was asked by Liu to recommend someone for the currency post in China, and suggested Dane. It would seem therefore that it was his Indian connections which gave Dane his second career in the East.[48]

At first sight, it cannot have seemed an attractive prospect. The German minister, it is true, believed from the outset that "the development of the salt administration will probably be like the maritime customs or the post office," and the Russian minister, too, argued that the chief inspectorate "is going to have very great influences upon China's finances." [49] British observers, however, were more sceptical. P. H. Kent wrote in August 1912, "With all respect to the great bankers . . . the administration of the Salt Gabelle under foreign supervision would present immense difficulties," and the *Times* gloomily commented on May 20, 1913, "It should be recognized by investors that in present circumstances the salt gabelle is practically worthless as a security." [50] Jordan himself in a despatch of March 31, 1913, in which he reported the Chinese intention to offer Dane the appointment, remarked that the Maritime Customs had not been very successful with their salt likin collectorates, and "it is questioned whether better success will attend the present effort." Langley, the assistant under-secretary, summed up the prevailing feeling with his minute on Jordan's despatch: "not very hopeful." [51]

It was, therefore, in an atmosphere of mixed expectations that the foreign salt gabelle got underway on May 21, 1913. It was launched against a background of deepening chaos in the traditional salt administration, a near bankrupt government in Peking, and the emergence of an articulate Chinese reform movement unsympathetic to foreign control. It was not an auspicious beginning.

4 | Dane's Reforms, 1913–1918: The Basic Achievement

The crucial phase in the modernization of the salt administration was the period of Sir Richard Dane's reforms, which followed the signing of the reorganization loan agreement. This chapter examines his four basic reforms: the establishment of the inspectorate, the centralization of the tax system, the campaign against tax avoidance, and the introduction of free trade. It begins by relating Dane's reform program to the concept of administrative modernization.

The meaning of modernization embraces concepts such as industrialization, mass participation in politics, administrative rationalization, increased social complexity. More abstractly, modernization could be defined, without violence to usage, as the process which the units of an institutional complex undergo when the number of units in the complex increases. An institutional complex might be population, or the politically involved part of a population, or the number of operations in an industrial activity.

This process has two characteristics, apparently antithetical but in fact complementary. On the one hand, the individual units in the growing complex become more versatile and better organized as the number of possible relationships they must adapt to continuously increases. Their potential efficacy is thus stepped up.

But at the same time, as the number of increasingly powerful units grows, the ability of any one of them to dominate the complex declines. Their relative predominance diminishes. The multiplication of the units of a complex, and the diversification of relationships this entails, thus tend both to increase and to diminish the power of the single units.

This model may be illustrated from the course of British administrative history in the nineteenth and early twentieth centuries, where, within an increasingly complex social and political context, an augmentation of power in some directions and a curtailment in others characterized the development of the bureaucracy.

On the one hand, the functions of the civil service expanded through being extended into those fields where private action was found to be ineffective. This was the policy represented by Sir Edwin Chadwick in the field of public health and Sir Robert Morant in that of education. On the other hand, equally as a result of the growing sophistication of society, the power of Whitehall was withdrawn from functions which could be more efficiently performed by other agencies or by the impersonal forces of the market. This was the line of the laisser-faire school, Sir James Stephen in the Colonial Office and Sir Charles Trevelyan at the Treasury. Bureaucratic power increased in the nineteenth century, but so did colonial autonomy and economic self-determination.[1]

Since Dane was a Victorian,[2] born in 1854 and passing the Indian civil service exams in 1872, it is not surprising that his reforms in the Chinese salt gabelle fell into this double pattern of modernization. This may be seen both from the way he approached the reform and from the actual measures he took.

Dane's philosophy of reform was set out in a memorandum he addressed to finance minister Liang Shih-i on June 24, 1913,[3] and in a note on the Ch'ang-lu salt division he wrote on August 24 of the same year. In the Ch'ang-lu note he argued: "the efficiency of a Salt Administration is determined by two tests: (1) the amount of revenue obtained by the Government, and (2) the cost of salt to the people. By both these tests the administration was condemned. The revenue was low considering the possibilities, while the prices at which the salt was sold to the public were high." [4] Dane, therefore, saw his task as twofold: to raise the yield and, at the same time, to reduce the rate of taxation. Wider power and increased efficiency would be needed to raise the total revenue of the gabelle, while, if the rate were cut, its hand would lie less heavily on the individual consumer.

To achieve this should not be difficult, he believed: "In a properly organised Salt Administration the interests of the Government and of the people are identical." Both would benefit from low prices and expanded sales: "The cheaper salt is the more people will use, and the more salt is used the larger will be the revenue." [5] Dane's picture of the future gabelle was of a single, national tax, kept as low as possible but levied universally and uniformly on a dynamic salt trade.

Despite the late Ch'ing and revolutionary reforms, the existing system was still "a complete negation of these principles," [6] and to bring the facts into line with the ideal, Dane saw four measures as essential. These too, it will be seen, fall into the pattern of expansion in one direction and contraction in another.

First, he must acquire effective power in the gabelle for himself and his subordinates. Second, the salt tax must be centralized, with the collections being made genuinely available to the central government alone and not siphoned off by local authorities. Third, the salt tax must be made comprehensive: all salt must bear taxation, not just the long-distance waterborne trade, as in 1913. All these reforms involved an increase in the power of the bureaucracy: more personnel, concentration of revenue, a wider basis of taxation.

But on the other hand, Dane believed that the salt trade must be expanded through competitive prices, competition itself to be a consequence of the introduction of free trade. This last reform would involve the partial withdrawal of traditional bureaucratic control in favor of reliance on the market.

Dane's reforms, therefore, fall into the pattern of modernization outlined above, and by 1918 when he retired from the gabelle, they had to a considerable extent been put into effect: there was a strong inspectorate, a centralized, comprehensive tax, and a large free trade area, though this last still left something to be desired.

The choice of Dane to undertake this reorganization might at first sight have seemed questionable. He was fifty-nine when he arrived in China in June 1913 and had spent the whole of his career in the Indian civil service in the palmiest days of empire, hardly the best introduction to the problems of a new republic; moreover, he knew no Chinese. He could be expected to be set in his ways, yet inexperienced in the peculiar problems of Chinese administration; it might have been argued that a senior customs commissioner — Konovaloff as suggested by the Russians or J. F. Oiesen as suggested by the Chinese minister — would have been a more appropriate appointee than this retired civil servant from the Indian civil service.

In fact, Dane's previous career fitted him well for his new job. The positions he had held in India had been mainly financial — commissioner of excise, secretary in the finance and commerce department — and he had ended his career as the first occupant of the post of inspector-general of excise and salt for the government of India. The generalist character of the Indian civil service, moreover, prevented his experience being limited to finance. He had served as assistant commissioner of the province of Ajmer, and, as boundary settlement officer in Central India in contact with the princely states, he had learned the elements of diplomacy. He was not without contact with Chinese affairs: he had served on the staff of the 1893–1895 royal commission on opium, submitting two memoranda[7] on the origins of the Opium war, and, as was noted in Chapter 2, he probably met the Chinese delegation which visited Simla in 1906.

Dane was a good choice from other points of view as well. An outdoor type of man, he was a noted traveler and sportsman. In China, this energy and stamina enabled him to visit all the salt divisions in person, which contributed immensely to the control he was able to achieve. As he put it, "To have shared the discomforts and hardships of an inspection tour in bad weather of salt works, which are usually in inaccessible localities, is in itself a tie; and it is much easier to understand and appreciate the soundness of principles when they have been personally explained and illustrated on the spot than when they have merely been inculcated in correspondence." [8]

At the same time, he was an expert fiscal technician who quickly mastered the intricacies of Chinese salt administration and who, according to Jordan, "talks salt morning, noon and night and has earned the sobriquet of the 'Salt Gabbler.' " His public personality was well caught by a perceptive correspondent of the magazine *Asia:* "His name somehow suggests his personality, shaggy and blustering with an immense capacity for concentrated work and for making other people work." [9]

Jordan's correspondence with Langley gives a less formal picture: "Dane is an impetuous Irishman who is impatient at all political difficulties and only wants to go straight ahead with his work . . . Dane, who is literally saturated with brine, has just come in to talk more salt . . . Dane stayed here a fortnight and sent in his resignation three or four times to mark his impatience of Chinese evasions and delays." To combine this kind of explosive approach with the patience and painstaking attention to detail needed for tax work argues an unusual blend of character. O. M. Green was right to believe that "his own personality was, of course, his greatest asset." [10]

The Establishment of the Sino-Foreign Inspectorate

The basis of all Dane's other reforms was the establishment of the Sino-foreign inspectorate. It was not only the first in time, but also the *sine qua non* of the others. Without effective power for himself and his subordinates, Dane could not have carried out his program at all.

The inspectorate, however, was not simply an operational base within the traditional gabelle. Its creation had a definite modernizing character, inasmuch as its establishment increased the power potential of the whole salt administration.

This increase was both quantitative and qualitative. There were now more officials, therefore the new inspectorate could undertake such tasks as the weighing of salt, which the old administration had either neglected altogether or performed inadequately. Once the inspectorate had the collection of tax entrusted to it, it acquired a large subordinate staff to control the scattered salt works and became involved in every aspect of the gabelle. In theory, control of the manufacture and marketing of salt, as well as of the salt police, remained in the hands of the preexisting administration. In practice, the members of the inspectorate assumed this control and more: they proposed reforms in salt mining in Yunnan, operated official salt markets in Szechwan, and in Fukien even conducted a primitive examination system for officers of the water police. The growth of the inspectorate, and indeed the scale of the gabelle, may be gauged from Table 11, which shows comparative

Table 11. Comparative costs for salaries and expenses in the salt administration, 1915–1925 (in Chinese dollars).

Agency	1915	1919	1922	1925
Traditional officials	1,675,100	1,796,607	1,902,601	1,930,164
Inspectorate	791,918	2,094,839	2,527,512	4,675,513
Salt administration as a whole	4,989,770	7,960,457	8,975,448	10,793,397

Source: Central Salt Administration Accounts.

costs for the traditional officials, the inspectorate, and the salt administration as a whole.

Even more significant, however, was the improved quality of administration which the inspectorate made possible. As a nationally organized body with minimum local ties, the inspectorate facilitated Dane's second reform, the centralization of tax; as a relatively incorruptible arm of government, thanks to its salary scale and modern ethos, it permitted the elimination of fees and perquisites; and its superior three-tier organizational structure ensured an all-round intensification of control.

Further, the inspectorate could extend the potential of the gabelle by the introduction of Western-style accountancy. Mathematical inexpertise hampered some Chinese officials: in October 1914, Dane complained that "Up to the present it has only been possible in the Chief Inspectorate to obtain the services of two Chinese officials whose Arithmetic can be relied on." Yet others were numerical virtuosos: Dane found that in Shantung, for example, "Duty rates were calculated to seven places of decimals." [11] What was lacking was any insistence on rigorous numerical definition, uniform for all the operations of the gabelle: the *yin,* for example, which was the whole basis of taxation, was not in itself a fixed quantity uniform throughout the country, though it was variably quantifiable at any given time or place. In this respect it resembled the tael and the picul. Toleration of the vague is a characteristic of premodern systems. Here again the inspectorate, with its standard unit of assessment, the *ssu-ma* picul, and its uniform dollar currency, was able to raise the overall level of efficiency.

Finally, the inspectorate, because of its specialized character and because it was new, energetic, and visibly effective, gave to the whole gabelle more esprit de corps and a higher morale than other government agencies possessed. This allowed the gabelle to continue operating despite the increasing political confusion of early republican China. In northern Szechwan, for example, in 1918, a year of peculiar disorder in this region, the inspectors reported: "our staff at the District Inspectorate, at the Collectorates and at the Kungyuans, displayed courage, sacrifice and loyalty to the Government. All employés under the control of this District Inspectorate, with very few exceptions, remained always, at risk of their life, at their posts, while all officials of other Government Departments, Civil Magistrates, Land-Tax Offices, Banks, etc. fled." [12] Purely administrative modernization had rubbed off naturally onto the individual's attitude towards his work.

Dane's program amounted in effect to a new salt law for the whole of China. Formally, the operational structure of the reformed gabelle was established by nine sets of regulations promulgated by the president of the republic during 1913 and 1914: regulations for standard units and rates (December 24, 1913), for the chief and district inspectorates (February 10, 1914), for the post of foreign adviser (February 20, 1914), for the manufacture of salt (March 4, 1914), for the central salt administration (May 15, 1914), for the deposit and transfer of salt revenue (May 16, 1914), against smuggling (December 22, 1914), and for the salt police (December 29, 1914). From these regulations, the inspectorate drew its authority.

By themselves, however, the regulations were only a legal schema. What enabled them to become operative was Dane's ability to exploit the crosscurrents of power he found about him — the Chinese government, the foreign interests, and his own position of relative neutrality between them — to secure compliance with his wishes.

In the first place, he could rely on the diplomatic support of the consortium bankers for whom the effectiveness of the inspectorate represented the security for their loan, and he was also backed by the ministers of the consortium powers whose prestige was involved in the success of the loan. Thus, in 1914, Dane worked closely with Jordan to secure the promulgation of the inspectorate regulations, which, although nominally accepted by premier Hsiung Hsi-ling as early as October 1913, showed no signs of being put into effect. The British government through Jordan made representations to Yüan Shih-k'ai, and in the following February Jordan commented on his negotiations with the Chinese authorities: "I have their definite assurance and mean to see that it is carried out." [13] Similarly in 1915, when the finance minister Chou Hsüeh-hsi threatened to throw over Dane's policy, Dane was able to invoke the support of the group banks.

On the other hand, Dane could mobilize Yüan Shih-k'ai's support for his plans by, on occasion, championing Chinese interests against the bankers. O. M. Green describes how Dane increased his prestige with the Chinese by insisting that the foreign banks pay interest on the salt revenues deposited with them. He adopted a similar attitude over the currency question. Postrevolutionary China, particularly Kwangtung, Szechwan, and Manchuria, was flooded with depreciating bank notes issued by local authorities. The consortium bankers disliked receiving salt revenue in the form of these notes and, in a letter of November 20, 1913, claimed the right to prescribe the currency in which the salt tax should be collected. From the Chinese point of view, this was objectionable: it would give the bankers the power to demonetize particular currencies at will, since refusal for tax purposes would be tantamount to repudiation.

On this issue, Dane sided with the Chinese and proposed a compromise which secured the Chinese objectives: the salt tax was first to be paid into Chinese banks in currency acceptable to the Chinese authorities and these banks would then convert it, at market rates, into currency acceptable to the consortium. This meant that the foreign bankers could not arbitrarily demonetize any currency they disliked or interfere with the Chinese government's rights to regulate internal commerce. Dane made it plain, however, that he expected his cooperative attitude to be reciprocated by the Chinese authorities.

Finally, Dane's position at the head of the inspectorate enabled him to influence the Chinese government through his control of considerable sums of money. On the one hand, there was the $20,000,000 from the reorganization loan, earmarked in annex F of the agreement for salt reform projects. Dane regarded most of these as unsound, but the money could not be applied to other purposes unless he gave his formal consent, and this he would not do, except on assurances from the Chinese government of acceptance of his policies.

On the other hand, there was the salt surplus, the net revenue when all liabilities had been met. In 1913, the prospects for such a surplus had been so remote that no provision for dealing with it had been included in the agreement. Dane used this silence to increase his own authority. He argued that since article 3 gave the chief inspectors the sole right of drawing on the account as well as the duty of safeguarding the obligations charged on it, they "become *ipso facto* the persons who determine the question whether the revenue collected is more than sufficient for the requirements of the obligations secured upon the salt revenue and whether there is or is not any surplus available for general purposes of the Chinese Government." [14] This made Dane, in effect, trustee of the salt revenue and increased his leverage with the Chinese government, the more so since, by 1919, salt was Peking's principal source of revenue.[15]

It was thanks to the skillful manipulation of this balance of forces that Dane and his successors were able to establish and maintain their position within the gabelle.

Pressure was most needed in the early days of the reform. Yüan Shih-k'ai's government was at first reluctant to implement the loan agreement. Dane described the situation in the provinces as follows: "The administration was conducted in the former manner [that is, as before 1912] by the Salt Commissioners . . . and the District Inspectors were mainly occupied in signing the receipts." In Peking too, Dane found obstruction at every turn: "Suitable stationery and ink were obtained with difficulty, and a telegram could not be despatched, except at private cost, unless it had been approved for issue by the Chief Commissioner or by the Minister of Finance." [16]

These obstacles were overcome by the tactics previously outlined. In August 1913, the chief inspectors obtained control of the salt revenue account: in accordance with the loan agreement, they alone could now draw on revenues once deposited with the banks. This, in turn, allowed Dane to exert control over the expenses and salaries of the administration, since revenue had to be deposited gross without any deduction. To make his control in this sphere effective, Dane demanded, and in September and October obtained, the power of audit over all salt offices.

In February 1914 came the decisive breakthrough: by new inspectorate regulations "which are practically his own handiwork," Dane secured for his subordinates the right to collect the tax itself and recognition of the principle that salt might leave the works only on the district inspector's release permit, showing that tax had been duly paid. As Jordan put it: "These Regulations will give Dane all the power he requires to place the Salt on a satisfactory basis and will strengthen his position immensely." [17]

The Centralization of the Salt Tax

The centralization of the salt tax was the most spectacular of Dane's reforms. Before 1911, the maximum salt revenue directly available to the central government was $20,000,000, the remaining two-thirds, $40,000,000, raised from the public in salt taxes being appropriated by local authorities or embezzled by private individuals. The change effected by Dane may be illustrated by Table 12, which indicates how greatly Peking's control over the revenue had been intensified. Even during the worst period of warlord confusion, 1920 to 1925, Peking's revenue remained substantially higher than it had been under the empire.

Centralization was accomplished by the direct duty system: that is, the collection of a single salt tax. Dane regarded this as the key element of his system: "The cardinal principle of successful salt administration is that the full duty . . . shall be levied before the salt leaves the works . . . once salt has been removed from the works on payment of the prescribed duty, the trade in it shall be facilitated in every possible way, and the salt shall not be liable to likin or any other form of local taxation." [18]

The traditional Chinese system (as has been shown in Chapter 1) differed from Dane's model in that there were multiple heads of assessment and multiple tax collections. For example, in 1913, under the old system, Huai-pei salt for north Anhwei paid eight taxes at Pan-p'u (the point of production), three at Hsi-pa near the Grand Canal (the point of first distribution), and four at Cheng-yang-kuan (the Huai valley general distributing center);

Table 12. Chinese salt revenues, 1913–1925 (in Chinese dollars).

Year	Net collections	Local withdrawals	Central receipts	Central releases
1913	11,471,242	–	11,471,242	–
1914	60,409,675	–	60,409,675	31,304,818
1915	69,277,536	498,804	68,778,732	27,024,256
1916	72,440,559	11,867,528	60,573,031	40,358,657
1917	70,627,249	7,496,942	63,130,307	61,116,428
1918	71,565,520	15,635,864	55,929,656	56,125,290
1919	80,606,503	26,341,208	54,265,295	48,842,241
1920	79,064,103	23,911,811	55,152,292	40,108,068
1921	77,987,838	18,414,094	59,573,744	52,060,237
1922	85,789,049	31,668,450	54,120,599	47,193,233
1923	–	–	–	–
1924	70,544,475	33,466,575	37,077,900	31,256,935
1925	73,634,425	33,029,660	40,604,765	32,935,633

Source: Central Salt Administration Accounts; Dane Report. Net collections are the total tax collected minus only the cost of collection; local withdrawals are that portion of the net collections spent by local authorities either with or without the consent of the central government; central receipts are the revenues directly available to Peking, being the net collections minus the local withdrawals; central releases are the revenues released from the central receipts to the central government, after the payment of obligations charged on the salt and the retention of a reserve for future liabilities.

while in Ho-tung, salt for Shensi paid eighteen taxes at the lake and a further 11 cash a catty on crossing the Yellow River at San-ho-k'ou.

Multiple collection could, it is true, be rendered compatible with centralization. On the Yangtze trade, Dane continued to make two levies: Huai-nan salt paid an advance duty at Yangchow of $1.50 a picul and a deferred duty of $3.00 a picul at the up-river port, while Ch'uan salt continued to pay both in Szechwan and at Ichang. In general, however, second collections facilitated decentralization and accordingly were abolished: for example, in Manchuria, in Ho-tung, in Huai-pei, and in the Shanghai region.

For multiple assessment, on the other hand, there was nothing to be said: it was inherently associated with the system of earmarking taxes for particular items of expenditure instead of paying them into a common treasury. Here, therefore, Dane had to make a complete break with the past, new regulations being issued on December 24, 1913: "the principle of taxation at the source . . . by the imposition of a single direct duty was definitely adopted; rates of duty were prescribed." [19] It was the collection of this tax by the inspectorate which produced the results shown in Table 12.

In bringing all the salt revenues under central control in this way for this first time, Dane solved a problem which had baffled Duke Tsai-tse and the late Ch'ing reformers. They had to operate with a decentralized bureaucracy in a decentralized situation. In the pre-Taiping bureaucracy, a man was appointed to his office by Peking but got his salary and necessary expenses from local sources governed by local conventions. Following the Taiping rebellion and the growth of regional power machines, the balance was tilted further in the direction of decentralization: a man's career, prospects, and even his ultimate political loyalty came to be centered on the governor-general. Officials thus enmeshed in a regional network of obligations and aspiration could not easily become the servants of a centralized government. In Franz Schurmann's terms, they had too many lateral connections.

Dane, on the other hand, employed a centralized administration in a centralizing political context. The inspectorate, partly because it was foreign, partly because it was new, could disregard these regional entanglements. Unlike the traditional official, the salt inspector, whether Chinese or foreign, drew all his salary and expenses according to fixed schedules laid down by the chief inspectors in Peking: the end of "corruption" thus automatically severed lateral ties. Unlike the traditional official too, the salt inspector could expect to spend all his working life in the gabelle. He served only one master and no longer had to keep an eye on the governor-general, the governor, or the censor as occasion arose. Finally, the salt inspector belonged to an organization which, from the start, thought nationally, in all-China terms: the problems of dynasty, culture, and nation which bedeviled the late Ch'ing reformers did not bother Dane.

Another asset, from Dane's point of view, for the centralization of tax was the development since 1900 of a nationwide banking system. The inspectorate provided officials who were willing to collect and deposit tax on behalf of the central government, but this would not have been enough if there had not also been a banking system capable of remitting funds and making them available to Peking. The Bank of China (1905) and the Bank of Communications (1907), which already had branches in most of the localities in which the salt gabelle operated, were ready instruments for this purpose. They were reinforced by the foreign consortium banks,[20] since the system of deposit and transfer devised by Dane and finally accepted by the bankers on May 16, 1914, was a dual one. Salt revenue was paid initially into the local branch of the Bank of China or the Bank of Communications, the so-called collecting banks. Next, the funds were transferred to a local branch of one of the consortium banks, designated as a receiving bank, which then again transferred them to the consortium head offices in Shanghai. Dane thus made use of an interlocking system of modern-style banks which covered the whole country. This may be contrasted with the more primitive customs bank system used by Hart in the nineteenth century, which possessed a lesser degree of integration and was more open to local pressures. Dane's deposit of salt revenue in turn strengthened the banking system and stimulated its development: for example, a special branch of the Bank of China was opened in the provincial town of Yün-ch'eng to receive the Ho-tung salt revenue.

The current of events favored centralization. The old regional power-holders had gone for good with the revolution. In 1913, most of the local authorities in China were comparatively new to office; many of them were operating new institutions such as military governments or self-government bureaus. They were, therefore, less habituated to controlling the salt revenues than the imperial governors-general, and they resented the intrusion of the inspectorate correspondingly less.

Yüan's consolidation worked in the same direction. The effective beginning of Dane's reforms in August and September 1913 coincided with Yüan's victory over the second revolution, which placed the government of most of central and south China in the hands of men who owed their position more to the president than to local support and who could therefore not afford to oppose him over salt policy.

For example, in the autumn of 1914, Feng Kuo-chang, the newly established Kiangsu tuchun, was offered 30 cents per bag of salt by the Huai-pei monopoly merchants to maintain their rights;

despite Feng, however, the monopoly was abolished by presidential mandate in December 1914. In Anhwei, Tuchun Ni Ssu-ch'ung, another beneficiary of the second revolution, established two companies to import Ch'ang-lu and Fengtien salt in January 1914. Such companies were a means of siphoning off taxes for local benefit, but, thanks to the support of the ministry of finance, Dane was able to secure their abolition in 1915.

In Kwangtung and Szechwan too, reform made little progress until the local leadership had been brought to recognize Yüan's authority. The pattern is confirmed by the cases of Yunnan and Kwangsi: here, the local tuchuns T'ang Chi-yao and Lu Jung-t'ing were never Yüan's men, and consequently the loan agreement was not properly applied. T'ang would allow the inspectorate to function only if he was paid a substantial subsidy out of the salt revenue, while Lu succeeded in preventing the establishment of the inspectorate until Yüan's political system collapsed in 1916. Kwangsi, however, was the only province except for Sinkiang where the loan agreement was not applied at all. Indeed, where Yüan's writ ran, and that meant most of central and south China, his authority was a significant asset to the reform.

A Comprehensive Salt Tax

Dane's third reform, the increase in the quantity of salt subjected to taxation, was the most exacting administratively. The degree of his success may be gauged by comparing Chang Chien's figures for taxed salt in 1911, themselves based on statistics compiled by the Ch'ing office of the controller-general of the salt gabelle, with the figures for salt released from 1914 through 1925, from the Central Salt Administration Accounts. The increase in taxed salt was not so striking as the increase in central government revenues, but it is clear that in its best years the inspectorate achieved an increase of between 30 and 50 percent in the quantity of salt subjected to tax.

This increase was partly the result of an expansion in the salt trade, which took place during the first decade of the inspectorate: more salt was flowing into the tax collector's net as a result both of Dane's measures to stimulate the salt trade and of the overall boom which the Chinese economy was experiencing.[21] But to a

Table 13. Salt subjected to taxation, 1911, 1914–1925.

Year	Quantity (piculs)
1911	26,867,936
1914	28,926,603
1915	28,754,619
1916	28,976,494
1917	30,739,553
1918	32,659,013
1919	36,057,323
1920	33,390,802
1921	36,528,550
1922	38,473,205
1923	35,942,490
1924	34,424,666
1925	33,817,985

Source: Chang Chien, *A Plan for the Reform of the National Salt Administration;* Central Salt Administration Accounts; District Reports, 1922.

greater extent, the increase in taxed salt was due to improved administration by the inspectorate, above all to more intensive administration. The traditional Chinese salt administration, like the rest of the Chinese government, was limited in range and superficial in its degree of control over the salt trade. The inspectorate, thanks to its superior organization and to its independence of local influence, was able to extend its control in three areas — production and storage, weighment and issue, and the control of transportation routes — and, *pari passu* with these developments, it was able to abolish tax farming.

Production and storage. "The principle of storage and control," [22] as the Ch'uan-nan inspectors described it in 1918, was a vital one for Dane's system. His plan for a single tax levied at source was similar to the *chiu-ch'ang cheng-shui* policy advocated by Chinese reformers from Ku Yen-wu to Chang Chien and Teng Hsiao-k'o. One of the objections already raised to this scheme was that it would be administratively impossible to exercise adequate control over salt works along the entire coast of China, or at the scattered wells of the interior. As Dane put it: "Control at the salt works of the production of salt was regarded as impracticable," and the traditional gabelle concentrated on controlling the outlets from the works along the main river routes. Dane's plan, set out in his memorandum of June 24, 1913, encountered the same objections: "Mr. Chi, the Chief of the Yen Wu Shu, pronounced the measures proposed impracticable, owing to the long coast line possessed by China." [23]

Dane did in fact experience difficulties on this score, and smuggling from the works was never eliminated: at one producing center in Szechwan in 1918, it was reckoned that 40 percent of the salt manufactured was smuggled out of it, while in Huai-pei in 1922, the inspector spoke of "the gigantic and systematic smuggling" prevalent in his area. Nevertheless, it is clear from the figures of salt taxed and from other local reports that effective control of the works was in most places achieved. For instance, at the principal producing center in Shantung, "The foundation of an efficient system for controlling the production, depôt storage, and weighment of salt was laid in 1914 and 1915 . . . It is safe to say that since 1915 not a single excess bag of salt has been transported by a merchant from these works." [24]

The apparatus of the gabelle was now brought closer to the salt works. In Szechwan, for example, the district inspector moved his headquarters from Lu-chou on the Yangtze to Tzu-liu-ching itself, while a new inspectorate (Ch'uan-pei) was set up in the north to control the scattered wells of that part of the province. In Ch'ang-lu, under the old regime, effective official control had commenced only at Tientsin: Dane moved it back to the merchants' depots at Han-ku, T'ang-ku, and Teng-ku. In Shantung, there was a similar move from Tsinan to the works themselves near the mouth of the Yellow River.[25]

Next, an attempt was made to ensure that all salt, as soon as produced, would be stored in officially supervised depots. These might be built and owned by the government. In Ch'ang-lu, for example, Dane built three depots, at Han-ku, T'ang-ku, and Teng-ku, at a cost of nearly $800,000; these had a capacity of over 15 million piculs, twice the Ch'ang-lu harvest in a normal year. Alternatively, the depots might be owned by the merchants and only supervised by the officials. This was the case at Tzu-liu-ching. Control, however, could be nonetheless effective. Here each store was to have one entrance and one exit, the key to the first being held by the manufacturers' representatives, the key to the second by the inspectorate releasing officers; all salt had to be deposited within forty-eight hours of manufacture. Government storage, however, remained the ultimate ideal: a small government store for 150,000 piculs was opened in the western part of Tzu-liu-ching in December 1920, and in 1921 an additional $345,000 was allocated for further building.[26] The inspectorate's aim was to bring all salt under control as soon as possible after manufacture.

A third technique used was the closure, subject to compensation, of outlying salt works which were too difficult to control. By regulations promulgated by the president on March 4, 1914, the gabelle acquired additional powers over the production of salt, which permitted the closing of inconvenient works. These powers belonged to the administrative branch rather than to the inspectorate, but Dane could act there too in his capacity of adviser,[27] and a number of works were in fact closed, notably in Fengtien, Ch'ang-lu, Shantung, Huai-pei, and Fukien, involving compensation of over $100,000. Closure was not extensively employed since the salt works were already relatively concentrated, but it made possible the closing of irritating gaps in the tax net.

Weighment and issue. Control of production and storage was only the first step. A more important checkpoint was what the inspectorate reports termed "weighment and issue," that is, the transfer from manufacturer to distributor, because it was at this point that tax was paid. The prime function of the inspectorate, in fact, was to issue release permits for salt against receipts of tax paid into the local Bank of China, and to ensure that no more salt was transported than had had tax paid on it. Physical control of the weighing out of salt at the depots was thus critical for the reform, and it was here that the incorruptibility and attention to detail of the inspectorate were important. Dane wrote in his report: "It was in the weighment and issue of salt . . . which Salt Officers in China had up to this time considered it unnecessary or derogatory to their dignity to attend to, that the assistance of the Foreigners employed in the Administration proved to be of the greatest and most immediate value." [28]

Various aspects of this assistance may be distinguished. By the regulations of December 24, 1913, the *ssu-ma* scale "was fixed as the standard scale for use in the Salt Department throughout China," [29] and while this was not rigidly enforced, it was adopted in most salt divisions. New weighing machines were accordingly made, either centrally by the ministry of finance or locally under the supervision of the district inspectors, and issued to the weighment officers. Under the old regime, salt had been taxed in effect by measure rather than by weight; this was now stopped. Another reform effected by Dane was the abolition, in February 1914,[30] of wastage allowances, that is, allowances of tax-free salt to compensate for possible losses en route of taxed salt. Admissible in theory, in practice these only complicated weighing and facilitated avoidance, and their abolition was another step towards making all salt sold from the works pay tax. In some divisions, notably the Shanghai area, standard bags for the packing of salt were also introduced.

Control of routes. The traditional gabelle, as we have seen, was levied in effect not on salt as such, but on salt traveling long distances along well-defined routes. Salt traveling either by land routes or short distances by water, or along inaccessible or unfamiliar water routes, went comparatively untaxed. Such salt, it is true, was a fraction of the whole, but it was a significant fraction, and one of the ways in which the inspectorate increased the total quantity of taxed salt was by bringing these routes under greater control.

The Pakhoi region of Kwangtung for example (including Hainan and the Lei-chou peninsula) was an important salt-producing area, whose production one estimate put at 1,600,000 piculs. Of this figure, 600,000 were exported overland into Kwangsi and then down the West River. The traditional gabelle paid little attention to this area, a quota of only 161,502 piculs being assigned to it under the Kwangtung salt laws, and the administration being rudimentary and largely farmed out to merchants. Dane therefore established a separate assistant inspectorate at Pakhoi with an appropriate staff, and, by 1919, 800,000 piculs a year were paying tax; in 1925, over a million piculs were paying tax. Here then we see the gabelle embracing a hitherto neglected overland route.[31]

In Kiangsu south of the Yangtze, the so-called Su Wu-shu was an example of an area supplied short distances by water, most of the salt required coming across Hangchow bay from Yü-yao near Shao-hsing or from T'ai-shan in the Chusan archipelago. The official quota of salt for this region, the most heavily urbanized in

China with a population of 12.5 million, was only 408,720 piculs, and while more was legally transportable, only 427,980 piculs paid tax in 1914. Dane established a new inspectorate at Shanghai (known as Sung-kiang district), cutting the administration loose from the Liang-che authorities at Hangchow, and, by 1919, 936,275 piculs were paying tax and by 1922 over a million were paying tax.[32]

An example of unfamiliar water routes coming under greater control may be obtained from the Shantung promontory. Here in 1913, "there was practically no administration at all," 30,000-odd taels being collected as a surcharge on the land tax on a nominal figure of some 68,000 piculs of salt. In fact, as the inspector noted, there was "a production of illicit salt amounting in one year to three or four million piculs, freely exported to Siberia and Hongkong, supplying three-fifths of the consumption of the entire country of Korea, and causing enormous loss of revenue to the Chinese Government in every maritime Province from Fengtien to Kwangtung." The struggles of the inspectorate to control this area, in which "the means of communication and transport are little better than execrable . . . and the track, such as it is, is more fitted for goats than for the passage of Government officers," were long and bitter and were not successful until after Dane left China. The planning, however, was done in his time, and an assistant inspectorate was created at Chefoo to conduct operations. By 1922 these plans had matured, and 1,482,100 piculs of promontory salt were being charged to tax.[33]

Abolition of tax farming. Tax farming, in the sense of handing over the entire administration of the salt monopoly to merchants in return for a lump sum, was a common practice of the traditional gabelle, particularly in areas such as Kwangtung or Kansu, where Peking's power was relatively weak. Substantial sums could be raised by this method,[34] but it involved an abdication of government responsibility and a consequent loss of control and revenue. Dane, therefore, eliminated the major areas of tax farming with beneficial results.

In the Swatow region, the so-called Ch'ao-ch'iao area, comprising districts in Kwangtung, Kiangsi, and Fukien, was in 1913 a comparatively well-organized part of the salt administration. The salt laws gave it a quota of 650,398 piculs and it was farmed out for

$700,000 a year. In 1916 the farm was terminated and an independent assistant inspectorate was set up to conduct the administration, though this was less of an innovation than at Pakhoi, since there had been a deputy salt commissioner stationed at Swatow in Ch'ing times. The Ch'ao-ch'iao inspectorate, like that at Chefoo, had numerous difficulties, since the Swatow region was peculiarly afflicted by political disorder. But already in 1917 it taxed 735,301 piculs of salt, and by 1922 it taxed 906,727 piculs and produced a revenue of $1,458,306. Tax farming even at its best could not do as well as direct administration by the inspectorate.[35]

Intensive administration, then, was the secret of Dane's ability to increase the amount of salt charged to tax. What gave the inspectorate this administrative potential? Incorruptibility and independence of the local scene are part of the answer, but another factor was its new method of organization. This may be described as a three-tier system: the inspector at the chief commercial center of the division where the bank was, the assistant inspector at the chief salt-producing area, and his subordinates at the individual

Table 14. List of assistant inspectorates, 1913–1918.

Division	Date	Location of assistant inspectorate
Fengtien	1913	Newchwang
Ch'ang-lu	1913	T'ang-ku
	1914	Shih-pei
Shantung	1914	Yang-chia-k'ou
	1917	Chefoo
Huai-pei	1916	Ch'ing-k'ou
Huai-nan	1914	Shih-erh-wei
	1914	Nanking (river forces)
	1915	T'ai-chou
Sung-kiang (Su Wu-shu)	1915	Liu-ho
Liang-che	1918	Ningpo
Kwangtung	1915	Ch'ao-ch'iao (Swatow)
	1915	P'ing-nan-kuei (Pakhoi)
Ch'uan-nan	1915	Wu-t'ung-ch'iao
Yunnan	1916	Mo-hei-ching
	1916	Pei-ching (Tali)

Source: Dane Report; District Reports, 1913–1917, 1918.

works. The traditional Chinese system by contrast was a two-tier one: there was no one of high rank between the salt commissioner in the divisional capital and the salt magistrates at the works. This meant that the salt commissioner's orders were inadequately transmitted and the salt magistrates inadequately supervised. The long lists of impeachments of corrupt salt officials which so frequently occur in the *shih-lu* for the last decade of the empire[36] are evidence of the inadequacy of the system in which surgery had to be resorted to instead of prevention. If this analysis is correct, then the assistant inspectors played a key role in Dane's system in extending the power of responsible officials one notch lower than the Ch'ing bureaucracy had done, thus making possible the extension of taxation.

Free Trade

The three reforms so far considered — the establishment of the inspectorate, the centralization of the tax, and the extension of the tax net — all involved an increase in the power of government. The last reform attempted, the liberalization of the salt trade, involved, on the contrary, a withdrawal of government power in favor of that of the open market. It might have been expected that administrative retreat would have proved easier than the administrative advances recounted earlier. In fact, free trade was a more radical break with Chinese tradition than the other reforms, because it implied structural changes in government and society beyond the confines of the salt administration. In the end, the foreign gabelle, specialized in function and circumscribed by the treaties, could not effect these changes, and consequently free trade, while it had its local successes, remained the least successful part of Dane's reforms.

He attached great importance to the principle. All the other reforms — taxation at source, consolidation of collections, audited expenses, and so forth — benefited the government; removal of restrictions to the salt trade would, he thought, carry reform to the people, to the producer and the consumer. Dane commented on the acceptance of his fiscal policies: "The reforms adopted were of immense importance, but they benefited the Government only. The important questions of transportation and sale were left entirely untouched, and the people were still left to the tender mercies of the monopolists." In his basic memorandum of June 24, 1913, he urged all-round liberalization: "The freest competition between different kinds of salt made at the different salt sources is desirable. Any attempt to regulate the areas in which the different kinds of salt shall be consumed (the Yinti system) cannot fail to be detrimental to the interests both of the Government and of consumers." Nor would revenue suffer if his views prevailed: "under a system of free trade the quantity of licit salt issued on payment of duty increases year by year with a corresponding increase in the amount of revenue received by the Government." Free trade then would make the reorganization popular and generate positive feedback to the revenue.[37]

Applied to Chinese conditions in 1913, free trade had a variety of meanings, all incorporated to a greater or lesser extent in the program.

In the first place, it meant the abolition of direct government monopoly: the purchase, transportation, and wholesaling of salt by the officials, the *kuan-yün kuan-hsiao* and *kuan-yün shang-hsiao* systems of the Chinese. Here Dane achieved a wide measure of success. Government transportation offices, as they were known, were done away with in the Honan section of Ch'ang-lu, in the north Anhwei section of Huai-pei, in Ho-tung where 33 out of the 120 districts had been administered in this way, in Jehol, Kalgan, and north Shansi along the Mongolian border, in the Su Wu-shu districts of Liang-che, in the Min valley area of Fukien, and along the frontier regions of Yunnan, to name only the most important.[38] By the end of 1918, Kirin–Heilungkiang, where Dane admitted conditions were exceptional, was the only sizable area in which government monopoly was still practiced.

Alternatively, free trade could mean the cancellation of monopolies of purchase, transportation, and sale granted to particular groups of privileged merchants: the removal of restrictions to participation in the salt trade. Here, Dane was less successful. In Szechwan, the monopoly of the waterborne trade, which was revived by Yüan Shih-k'ai's government in 1915, was successfully abolished in 1916; in Huai-pei, the privileges of the *p'iao-shang* who conducted the trade from the works to Hsi-pa were cancelled; in Liang-che, the monopolies at Shao-hsing, T'ai-chou, and Wenchow were eventually abolished; in Kansu, proclamations announcing the introduction of free trade were issued in Chinese and Tibetan on November 1, 1918; while in Kwangtung, monopoly in the Swatow (Ch'ao-ch'iao assistant inspectorate) and Pakhoi (P'ing-nan-kuei assistant inspectorate) regions disappeared soon after the tax farming system went out in 1916, and in 1917 there were over a hundred merchants in each area engaged in free trade.[39]

On the other hand, Dane failed to persuade the ministry of finance to abolish monopoly in Ch'ang-lu. In Sung-kiang inspectorate in 1922, the salt business was still dominated by an old-style consortium, the Su Wu-shu merchants' association: "The activities of the Association include direct dealing with the Salt Administration, purchase of salt from works, transportation of salt to depot, supervising the working of their depots in the works, at Liuho and Yeh-hsieh, and application for permits and passes, as well as payment of duty." [40] In Szechwan, in 1918, the inspector reported that the Kweichow trade "is largely, if not entirely, controlled by the transporting merchants who held the monopolistic rights during 1915–16." [41] Furthermore, Dane did not upset the arrangements in Szechwan for monopoly markets (*kung-yüan*) for land-carried salt, which were introduced in 1915 at the same time as the revived monopolies for waterborne salt. In 1917 the Ch'uan-pei inspector even praised them as indispensable aids to administration: "If it were not for the 158 Kungyuans that are established in Chuanpei how could we collect taxes from the 7,585 salt factories existing in this district and scattered in one area of 100,000 square *li?*" [42] Dane's original program for introducing free trade in Huai-nan division was indefinitely shelved, while no attempt seems even to have been made to abolish monopoly in Shantung and Ho-tung. The map of China in 1919 remained a patchwork of free and restricted trade areas.

Yet again, free trade could mean the lifting of restrictions on the source of supply and on the choice of market, on the quantities allowed to be transported, and on the price of salt. These might or might not go hand in hand with the other kinds of free trade.

In Ch'ang-lu, for example, Dane failed to break the merchants' monopoly, but on the other hand salt was permitted to be sold freely in certain parts of the neighboring Huai-pei, Ho-tung, and

Chin-pei divisions. In the water-supplied regions of Szechwan however, after merchant monopoly was abolished in 1916, the producing areas retained their particular consumption areas, though these were not stringently enforced and the inspectors refused to accept any legal limitation on the quantity of salt sold. In April 1914, limitations on the number of licenses issuable in a year, and hence on the quantity of salt in circulation, were abolished in Shantung, and in Sung-kiang inspectorate too, from 1915, "no limit was set to the quantity of salt transported by the merchants," [43] though both these areas continued to be dominated by merchant monopolies. In Kwangtung, however, by 1917 freedom was theoretically complete: "After registration and the issue of transportation passes on payment of duty, the merchants have a free hand in their business without interference on the part of the administration. No arrangements regarding the price of salt were fixed by the Government. The prices of salt depend entirely on the quality of and demand for salt." [44]

There is evidence that greater freedom for market forces did expand salt sales and reduce salt prices. This comes out clearly in the case of the Szechwan waterborne trade. In the twelve months before the establishment of the monopolies on June 1, 1915, the three main producing centers exported 4,109,547 piculs of duty-paid salt. Under the monopoly the same three centers exported 2,757,780 piculs between June 1, 1915, and May 31, 1916, while in the year following the abolition of the monopoly, free trade merchants handled 3,753,180 piculs of salt. The monopolists could argue, it is true, that their trade had been disorganized by the revolt against Yüan Shih-k'ai in the early part of 1916, but the summer of 1917 saw civil war in Szechwan, and anyway it was part of Dane's argument against restriction that a single monopolist was more vulnerable to political disorder than a group of lesser merchants.[45]

In other areas too, the inspectors attributed increases in the release of salt or falls in its price to the effects of liberalization. In 1918, for example, issues of salt from the Ch'ang-lu depots increased by 1,600,000 piculs, "this increase being principally due to the 'borrowing' of Changlu salt by the Hupei and Anhui merchants and the opening of the Chinpei nine, and the Honan eight, districts to Changlu salt." In Huai-pei in 1921, "Owing to the abolition of Yin rights the release for North Anhui reached a record figure of 1,831,110 piculs or an increase of 119,960 over last year." In Kansu in 1918, "As a result of competition under free trade, the retail price of salt has everywhere dropped considerably; for instance, in Lanchow a catty of salt, which cost formerly 112 cash, can be had now for 78 cash." In such cases, other factors may have been at work and something should be discounted for the inspectors' enthusiasm, but these examples do provide prima facie evidence for the success of Dane's system.[46]

Why then was it so strongly resisted, and even in areas where nominally introduced — the Szechwan Kweichow trade referred to above — unable to take root? The principal reason was that the argument of ready money, which Dane used so effectively to get his other reforms accepted, could here be used against him. The governments of the early republic, both national and provincial, lived from hand to mouth, and immediate benevolences from merchants, usually in the form of contributions to loans, might appear more attractive than the prospects of increased salt revenue through an expansion of trade. Furthermore, such benevolences, which were always forthcoming in return for privileges, were a means of milking the salt revenues outside the normal channels of the gabelle.

Dane was aware of this situation: "Rumours of an intended levy of a contribution of $700,000 from salt merchants in the Changlu Yinti area had reached the Foreign Chief Inspector . . . In return for a forced payment or a voluntary contribution by salt merchants to provide for some immediate necessity, it was customary in China to sacrifice the interests of consumers and also, it may be added, millions of annually recurring revenue." [47] Again, in connection with the proposed reestablishment of monopoly companies in Szechwan, he wrote in a memorandum of November 1914: "If, (as appears probable), these measures are being taken in connection with the issue of bonds of the Domestic Loan, the unwisdom of the course followed is even greater. For the sake of a few hundred thousand dollars, which have to be repaid, the Government is sacrificing millions of direct net revenue." [48]

He had, however, to recognize the strength of the forces opposing him: he could defeat particular proposals, but he could not correct the bias of the system towards monopoly and government control because, in the last resort, of the inadequacies of Chinese public finance in the early republic. Thus in Ch'ang-lu, the proposal for a benevolence was dropped, in form at any rate, but the minister of finance refused to introduce genuine free trade.

Ch'ang-lu was in many ways the best ordered salt division, the one most under Peking's control. Dane recognized the significance of his defeat here in a marginal note: "Hope of effecting any general improvement in the arrangements for the transportation and sale of salt throughout China abandoned in consequence of this episode." [49]

Dane exaggerated his defeat: significant liberalization of the salt trade was achieved in Szechwan, Kwangtung, Huai-pei, Shanghai, Chekiang, Yunnan, Fukien, and much of the northwest. Nevertheless, the failure to establish a comprehensive system of free trade marked the limit of his reforms. Beyond this point his power to achieve significant change ran out.

Dane's reforms have been described from the standpoint of the definition of modernization put forward at the opening of this chapter, that is, in terms of the parallel increase and diminution of function that bureaucratic development within a modernized complex was held to involve. At this date, however, China was only partly modernized. It was in fact the extent to which the social and political complex remained unaffected by modernization that delimited the area of Dane's achievement.

His success in China, so far as it went, had been due to an appropriate combination of aims and methods: on the one hand to his skill as a fiscal technician, and on the other to his confining his attention to objectives which were within the scope of the power he had at his command.

As a European fiscal expert, with a century of economic science and half a century of Gladstonian finance behind him, Dane had a clearer sense of priorities and of the relation of means to ends than his Chinese contemporaries had. Purer salt, equalization of salt prices, improvement of the lot of the salt producer — these were desirable aims, but they were not allowed to distract from "the primary object of the reorganisation, namely the production of revenue." [50] Traditional thinking failed to isolate fiscal theory from political morality and from administrative detail, and consequently found itself faced with a plurality of goals and a confusion of techniques. Dane, on the other hand, could take a single aim and direct all his measures to that end.

Dane's primary aim, the maximization of central government revenue, was to be achieved through a single tax levied at source on all salt. This involved both a centralization of revenue previously dispersed and an intensification of the control of the gabelle over the salt trade. Both measures were well within the range of the kind of administrative action open to Dane; indeed, they were peculiarly suited to it.

As a foreigner and a newcomer to China, Dane's position within the Chinese salt administration was inevitably that of an intruder from outside. This both limited and facilitated his freedom of action. On the one hand, he lacked knowledge of the system, he had no experience of its working, and he enjoyed no automatic support from any body of opinion. On the other hand, unlike reformers from within, he had no previous commitment to policies or personalities: he could treat the problem of reform in abstract, on a priori terms, and frame a program unencumbered by the legacy of Liu Yen, T'ao Chu, and Ting Pao-chen. Furthermore, the measures he ultimately proposed — a new administrative hierarchy, centralization of tax, extended control over storage, issue, and routes — lent themselves to the authority he exercised, which was itself centralized and operated vertically from above to cut through the old autonomies and local idiosyncrasies of the traditional gabelle.

To generalize, it may be argued that the range of action open to the foreign administrator in China was the exact opposite of that which lay before the Chinese reformer working from within the system. Where the administrative task was technical, specialized, and relatively insulated from politics, reform from outside was the more effective, permitting a concentration on essentials and an overriding of vested interests. Dane's reform falls into this pattern, as does Sir Frederick Leith-Ross's currency reform in 1935. Where, however, there was less specialization and political questions were deeply involved, the foreign administrator was less effective: he could not amass support, exert pressure, or control the invisible determinants of the situation. In such situations it was no help to try to cut the Gordian knot. This was the weakness of the German military mission in the 1930's,[51] and Dane himself would have run into similar difficulties if anything had come of the suggestion, made in 1915,[52] that he go on to reform the land tax. It seems that synarchy to be successful had to concentrate on narrowly defined and politically neutral aims.

Dane's failure over free trade is, therefore, not surprising. For here, he was leaving the confines of fiscal administration and entering a field with extensive political and economic ramifications, where modernization had as yet made few inroads. An effective system of free trade would have involved a readjustment of China's system of public finance at all levels, so that officials were no longer dependent on irregular exactions from salt monopolists. To construct an open market to which the regulation of trade in salt could confidently be transferred was a task far beyond the unaided capacity of the foreign gabelle, circumscribed as it was by its specialized function, the reorganization loan agreement, and the general character of the European presence under the treaty system. Here Dane was in the position of a Seeckt or a Stilwell: he saw what should be done but was unable to manipulate the recalcitrant Chinese environment.

5 | Dane's Reforms, 1913–1918: Reform in the Salt Divisions

So far, Dane's reforms have been considered as a whole: this was how he himself saw them, as constituting a single system. However, this was only half the story: to obtain a complete picture of the reforms, one must also look at how they were applied in the particular salt divisions, each with its own difficulties and opportunities.

Dane himself was aware of the diversity of conditions with which he had to deal. "To attempt to upset all these different ways of practice and to establish some cast iron system in their place," he stated in an interview with the *North China Daily News* on April 6, 1914, "would cause infinitely more harm than good." [1] In his memorandum of June 24, 1913, he argued that "in introducing reforms it is very important to make as little alteration as possible in the conditions, to which the people are accustomed," subsequently claiming that he proposed changes only where existing practice was plainly incompatible with the objectives of the reform.[2]

Of Dane's four basic reforms — establishment of the inspectorate, centralization of tax, extension of the tax net, introduction of free trade — it was the last two which varied most in application. The inspectorate was established in all the salt divisions, and in all of them the tax collected was brought under the control of the central government. But the extension of the tax net came up against different problems in different areas: in some, it was a question of improving control over routes already controlled; in others, it was a question of bringing new routes, either water or land, under control. As regards free trade, Dane was in some cases unable to put his program into practice at all; in others, he deliberately modified it to compromise with local circumstances.

As a standard of comparison, figures for taxed salt are given in this chapter (Tables 15–18) for each of the twelve salt divisions.[3] These are the best index for the effectiveness of reform since, unlike tax receipt figures, they are not affected by inflation or changes of tax rate. Dane himself left China in November 1918, but since many of his reforms began to show results only after this date, the tables have been extended into the early 1920's and thus cover the first decade of the inspectorate's operations. The taxed salt statistics have another advantage in that they reflect two distinct factors: on the one hand, the inspectorate's degree of control over the salt trade and, on the other, the state of the trade itself. Thus in any particular area, an increase in the amount of salt taxed may be due either to better control or to an expansion in trade, and it is usually possible by closer analysis of the figures to decide which, as, for example, with Huai-nan, Manchuria, and Liang-che. Dane, of course, wanted both: improved control was his immediate objective, but in the long term expansion of trade was what he depended on to produce a continuous increase in revenue.

In the analysis of differences, use will be made of the geographical schema of the salt trade presented in Chapter 1: a major east-west axis along the Yangtze, the minor north-south axis from Manchuria through Peking to Canton, together with the quadrants, east and west, between the two axes. In terms of this schema, Dane's impact was strongest along the minor north-south axis, where modern communications favored reform, and in the eastern quadrants, where the inadequacies of the Chinese institutions made it most necessary. His impact was weakest along the major east-west axis, where Chinese institutions were at their best, and in the far west, where primitive conditions hindered Chinese and foreign reform alike. The failure of free trade in Ch'ang-lu and its surprising success in Szechwan are exceptions to this generalization.

CHAPTER 5

The Major East-West Axis

Along this axis were situated the two premier salt divisions in terms both of revenue and of salt handled, Huai-nan and Szechwan. Though these two regions were similar in national importance and in the sophistication of their traditional institutions, the impact of Dane's reforms in them contrasted sharply.

Huai-nan. It might have been expected that Dane would here have advocated the most drastic reforms, since this division exemplified the principles of the old order in their most developed form. In fact, the opposite was true: "Sir Richard Dane, when he inspected Lianghuai, was so impressed with the potentiality of the system, so far as the collection of revenue in the four Provinces was concerned, that he did not recommend any material change in it." [4]

Dane's conservatism in Huai-nan is most evident in the matter of free trade. Though he disliked the high prices and heavy tax produced by the monopoly system, he believed that the first priority must be given to restarting the system after the confusions of the revolution: "For the present, the establishment of a financial equilibrium is an absolute necessity," and this involved temporary acceptance of the traditional arrangements. Moreover, he regarded the Yangtze monopolists in a different light from those of other areas, giving a remarkably traditional reason for his discrimination: "It is also only fair to the Yangtze Yin merchants to say that, although they make large profits, they do render to Government some service in return, as they bear the whole cost of the purchase and transportation of salt for sale under the Government monopoly. In this respect the Yin rights in the Yangtze differ materially from the similar rights in Huaipei and in the Changlu area." [5] Like all Liang-huai reformers, Dane balked at the task of finding sufficient government capital to handle the Yangtze trade.

In spite of this reluctance to interfere with the commercial mechanisms of traditional Huai-nan, Dane still made a number of changes to make the tax net more effective.

On the one hand, he improved control over the main artery of the system, the route from Yangchow to the up-river ports of

Anhwei, Kiangsi, Hupei, and Hunan. He began with the depots: "the cornerstone of the edifice," as Dane called it, was the salt depot at Shih-erh-wei, up the Yangtze from Yangchow, where the salt passed from the producer (*ch'ang-shang*) to the transporting merchant (*yün-shang*). Without proper control over this depot, no realistic assessment of tax was possible. In 1913, official control was lax in the extreme: "When Sir Richard Dane visited the Depot in December 1913 he found that the Officer, who was nominally in charge, had not been at Shiherhwei since the month of July, and was residing at Shanghai or elsewhere. The management of the Store was practically entirely in the hands of the merchants." [6] Dane changed all this: an assistant inspectorate was set up at Shih-erh-wei, the Chinese staff at the store were registered and given badges to identify them and prevent petty theft, and a European was placed in charge of the river police.[7]

Second, the amount of tax payable at Yangchow, as opposed to the up-river port, was substantially increased. In 1913, salt for Hupei, for example, paid less than 2 taels a *yin* at Yangchow or Shih-erh-wei and over 18 on arrival at Hankow. The long river journey facilitated avoidance; in 1914, therefore, Dane altered the system to $1.50 a picul payable at Yangchow and $3.00 at the up-river port. This move was not popular with the merchants since the larger advance payment increased the amount of capital required for the business, and they organized a strike against it from October 1914 to January 1915. The government's determination, however, forced them to give way.[8]

Finally, control was strengthened in the up-river provinces, where the bulk of the Huai-nan taxes were still collected, at the point when salt passed, under official supervision, from the transporting merchant via the provincial salt hong to the local dealer. In July 1914, an inspectorate audit office was set up in Hankow to scrutinize these transactions and to make sure that the salt sold tallied with that issued at Shih-erh-wei; further offices were later established at Changsha, Nanchang, and Wuhu. As at the outset of the reform there had been some doubt whether the loan agreement covered these inland taxes far from the source of supply, a further aim of the audit office was to universalize inspectorate control. As Dane put it: "It was in fact probably the most conspicuous outward and visible sign of the policy . . . under which the whole of the salt revenue throughout China, from whatever source it was derived, was brought to audit." [9]

Besides increasing control over the main long-distance water route in Huai-nan, Dane also achieved an improvement in the taxation of the *shih-an,* or local consumption areas. This rich urbanized region, which included Nanking, Yangchow, and cities along the Grand Canal, was a typical underassessed area of the old regime: it lay close to the salt works and was supplied by a complex network of short-distance water routes. At the time of the customs report, its quota of taxed salt was only 381,360 piculs,[10] though the true consumption was nearer a million. Thanks to the assistant inspectorates at T'ai-chou which supplied northern Kiangsu and at Shih-erh-wei which supplied Nanking, taxed salt for the *shih-an* increased steadily from 957,208 piculs in 1916 to 1,490,976 piculs in 1922.[11]

The figures for taxed salt given in Table 15 indicate the extent

Table 15. Salt releases in Huai-nan and Szechwan, 1913–1925 (piculs).

Year	Huai-nan	Szechwan
Prereform	4,131,360	5,107,120
1914	–	3,502,081
1915	4,377,198	5,361,131
1916	4,630,798	6,388,885
1917	3,603,331	6,101,916
1918	4,604,896	6,132,219
1919	4,921,247	6,105,763
1920	4,319,602	6,391,262
1921	5,068,645	6,058,832
1922	4,898,558	6,423,239
1923	4,521,127	6,133,912
1924	4,432,233	6,541,555
1925	5,161,809	5,453,602

Source: Customs Report; Dane Report; District Reports; Central Salt Administration Accounts.

of Dane's success. The increase in taxed salt was due to the rationalization of control of the *shih-an,* not to any increase in up-river trade. Dane could not achieve any expansion over the 1905 total in the salt issued at Shih-erh-wei for the up-river ports:[12]

the retention of the monopolistic framework handicapped him severely in his assault on this, the longest and richest of the water routes.

Szechwan. Here, on the contrary, surprisingly for a province not firmly controlled by the central government, Dane's reforms left a deep mark on the institutions of the salt division, in particular his partition of the area into the two inspectorates of Ch'uan-nan and Ch'uan-pei.

The Szechwan salt administration, essentially the work of Ting Pao-chen, the last traditional salt reformer, was, in Dane's view, "probably the best in China." [13] Nevertheless, in terms of the effectiveness of its control over the salt trade, it still left something to be desired from a modern standpoint. Dane estimated the total production of the division at over 7,000,000 piculs, and of this only 5,107,120 were brought under taxation.[14]

The waterborne or *yin* trade, the traditional staple of this salt division, was reasonably efficient: "In the 78 districts of Szechuan, in Kweichow and in the portions of Yunnan and Hupeh, which were supplied with Szechuan salt by transportation on the Min and Yangtze river . . . the administration for control of the transportation was highly organised." [15] Even here, however, there were defects in 1913: "No official control whatever was exercised over production," and "It was only on arrival at the first Checking Station that the salt came under official supervision." [16] The foreign gabelle changed this by moving the inspectorate from Lu-chou on the Yangtze to Tzu-liu-ching itself and by establishing an assistant inspectorate at Wu-t'ung-ch'iao to supervise the Chien-wei and Lo-shan group of wells on the Min River below Chia-ting. By 1917, with the appointment of inspectorate collectors to the Yün-yang and Ta-ning wells in the Yangtze gorge region, "the release and weighment of all waterborne salt throughout Chuannan was brought under effective control by staffs appointed by and sub-ordinate to the District Inspectors." [17]

This, however, was only a beginning. In 1918, as described in Chapter 4, government-supervised depots were introduced at Tzu-liu-ching for the storage of *yin* salt immediately after production, and the following year the system was extended to the next most important producing centers, Chien-wei and Lo-shan. *Yin* salt, then, was well controlled by the end of Dane's term of office.

A more serious challenge to the principle of a comprehensive salt tax was posed by the *p'iao* or land-carried trade in Szechwan. Dane wrote: "In the remaining 68 districts in Szechuan, to which salt was transported by land either by porters or by pack animals, the taxes were very light and there was no effective administration." [18] The basic problem was that while only seven producing centers were involved in the *yin* business and the trade was thus bound to follow clearly defined routes, in the *p'iao* trade there were twenty-six producing centers with routes wherever an animal or man could walk. The problem was particularly serious in the mountainous northern part of the province where the wells were scattered and law enforcement of any kind difficult.

Before the reform, only 1,375,000 piculs of *p'iao* salt were subject to taxation. By 1918, this had been increased to 2,655,377 piculs.[19] The measures already outlined for increasing inspectoral control at the chief *yin*-producing towns were partly responsible, since these centers also produced salt for the *p'iao* trade. In addition, the *kung-yüan* system of marketing, to be discussed below, facilitated taxation by confining the trade to fewer channels. The most important measure, however, for controlling the *p'iao* trade was the establishment in June 1915 of a separate district inspectorate for northern Szechwan (Ch'uan-pei) with headquarters at San-t'ai on the Fu River, a tributary of the Kialing. In 1918, it brought 1,377,148 piculs of *p'iao* salt under taxation, more than the total for the entire division under the old regime. "Nearly all the salt in that area is P'iao or land-borne," [20] and the new inspectorate, with satellite collectorates at the outlying wells, was thus bringing the gabelle into a region where, before the reform, there had been no effective control at all.

Dane considerably extended the tax net in Szechwan; he was also active in promoting the salt trade through changes in commercial organization. As regards the *yin* trade, his chief measure was the reestablishment of free trade on September 8, 1916. Free trade had been introduced by Teng Hsiao-k'o during the revolution of 1911, when it replaced Ting Pao-chen's government monopoly. In June 1915, against Dane's strenuous opposition and largely because of the advocacy of Yen An-lan, former head of the salt administration under Duke Tsai-tse and now Szechwan salt commissioner, monopoly — this time merchant monopoly —

was reestablished. Dane, however, continued to oppose it, backed by E. von Strauch, the foreign deputy chief inspector who, as former Chungking customs commissioner, had a special knowledge of local conditions. In September 1916, the companies' privileges were canceled and the Szechwan *yin* region became the largest free trade area established by Dane.[21]

As regards the *p'iao* trade, however, Dane adopted a different policy, significant as an attempt to adapt rather than abolish the principle of monopoly so deeply rooted in Chinese conditions. Until 1915, trade in *p'iao* salt had been entirely unrestricted: in that year, as part of Yen An-lan's monopoly scheme, 202 *kung-yüan* (lit., "public depots") were established. These were privately owned stores, which were given the sole privilege of buying from the producer and selling to the wholesaler or retailer. They were, in fact, monopoly marketing agencies, their advantage to the government lying in their canalization of the trade.

The disadvantage was to the small producer, the rising capitalist, since the majority of *kung-yüan,* particularly at Tzu-liu-ching, were owned and operated by the large manufacturers, who saw to the sale of their own salt first. To counter this, Dane sanctioned in 1917 the setting up of *kuan-yüan* (lit., "government depots"). The Ch'uan-nan inspectors described their function as follows: "These establishments will be operated in conjunction, and in competition, with the present merchants' Kungyuan and will give all p'iao salt producers who are not shareholders in the present Kungyuan an opportunity of disposing of their salt." The experiment, which broke new ground for the inspectorate, was a success, particularly as regards promoting trade. The Ch'uan-nan inspectors commented in 1922: "The establishment of Kuanyuan, which were allowed to trade side by side with the Kungyuan, produced a greatly stimulating effect on the Piao salt trade." [22]

The new marketing arrangements for *yin* and *p'iao* salt as well as the improved control of the routes are reflected in the figures for taxed salt (Table 15), which show an increase of a million piculs over the prereform figures. Most of the increase was due to the *p'iao* trade, since the *yin* figures never passed the prereform total. But by the end of Dane's term of office, the *yin* figures too would have been higher if it had not been for the political disorders that upset Tzu-liu-ching's trade with Hupei and Hunan.

In a normal year, the export to Hupei and Hunan amounted to over 1,200,000 piculs, but in 1918 interference by the Hupei tuchun reduced it to 625,000 piculs, and in 1920 it fell as low as 529,200 piculs.[23]

The Minor North-South Axis

Here, from north to south, were situated the salt divisions of Manchuria, Ch'ang-lu, and Kwangtung. These were regions of maximal impact for the inspectorate.

Manchuria. Of these three regions, Dane's influence was least felt in Manchuria. Prior to the revolution of 1911, the Manchurian salt trade had already moved onto a modern communications basis, using the South Manchuria railway for Fengtien and Kirin, and the steamer to Vladivostok and thence the Chinese eastern railway for Heilungkiang. These routes were already adequately controlled by the institutions set up by Hsü Shih-ch'ang in 1908: a line of eight likin offices at the chief producing centers in Fengtien and government transportation of all salt for Kirin and Heilungkiang.

Two problems, however, remained. The first was the extent of the salt fields in Fengtien. Most of the salt consumed in Manchuria came from Ying-k'ai or Fu-hsien to the east of Newchwang and was easily supervised by the district inspectorate at Newchwang. Works, however, existed all the way from Shanhaikuan to the Yalu, producing innumerable local transportation routes and making taxation at source difficult. The only solution was closure against compensation. As the Fengtien inspectors put it in 1919: "To enclose in the smallest possible area the innumerable salt producing marshes of this district has been our constant policy since the reorganization of the Salt Administration." There were limits, however, to the possibilities of closure, and the problem of the Fengtien coastline was never entirely solved. As Dane admitted, the division "is not an easy one to administer." [24]

Even more intractable was the problem of smuggling from the Japanese-leased territories. Here there was a booming salt industry (420,000 piculs in 1907, 1,687,623 in 1913) ostensibly directed at Korea and Japan, but salt was frequently smuggled with Japanese connivance up the South Manchuria railway into Chinese territory.

In 1916, the Fengtien inspectors noted: "During the year this office received two petitions from the Fengtien (Mukden) Salt Trade Association appealing for help in combating the influx of Japanese salt, which was ruining their trade." [25] In 1917 an agreement was signed with the South Manchuria railway authorities, introducing railway passes, but the problem was never definitively solved.

As regards marketing arrangements, Fengtien in 1913 was open to free trade (no monopolies had ever existed), while salt for Kirin and Heilungkiang was handled by government transportation which the salt commissioner planned to extend to Fengtien also. Dane decided to maintain the status quo: Fengtien's free trade was in line with his system, while the special circumstances of Kirin and Heilungkiang as pioneer provinces with well-defined communications justified the retention of official transportation. He consolidated the two provincial offices into one (Ki-hei) with headquarters at Ch'ang-ch'un and appointed a foreign auditor, William Morgan Palmer, with functions similar to those of the Yangtze auditor in Huai-nan. A number of useful reforms — abolition of wastage allowances, public tenders for salt at Newchwang, greater freedom of consumer purchase — subsequently increased the efficiency of the monopoly.[26]

Manchuria experienced a sharp falling off in salt releases (Table 16) in the early years of the inspectorate, thanks to an unwise increase in regional tax rates in 1915 which Dane had unsuccessfully opposed.[27] The natural dynamism of the region eventually overcame this handicap, and by the end of the decade Manchuria was handling 700,000 more piculs than before the reform. Most of this progress was attributable to Kirin–Heilungkiang: before the reform the monopoly handled 900,000 piculs; in 1922, it handled 1,526,256, while in 1919 the figure had reached nearly 2 million.[28] Population was rising rapidly in the two northern provinces, with Harbin in this decade becoming the largest city in Manchuria as the importing center for Russia's wartime supplies. The monopoly made its own contribution to the progress of the Manchurian cities, with the auditors for Ki-hei noting in 1919 that "This large salt trade materially added to the prosperity of the city of Changchun." [29] Urban development and a growing market for salt thus provided ideal conditions for Dane's system

Table 16. Salt releases in Manchuria, Ch'ang-lu, and Kwangtung, 1913–1925 (piculs).

Year	Manchuria	Ch'ang-lu	Kwangtung
Prereform	3,840,000	3,974,982	1,954,821
1914	4,206,391	5,794,605	2,606,796
1915	2,494,037	5,152,560	2,179,872
1916	2,527,966	4,551,384	2,018,502
1917	3,290,698	4,092,964	4,516,809
1918	3,237,156	5,466,272	4,048,331
1919	4,544,649	5,287,468	3,398,018
1920	3,404,718	4,743,141	3,315,937
1921	4,381,137	5,005,172	4,362,020
1922	4,551,115	5,133,076	4,496,084
1923	4,162,149	5,621,652	–
1924	4,738,191	4,577,270	2,072,487
1925	4,477,705	4,741,111	–

Source: Chang Chien, *A Plan for the Reform of the National Salt Administration;* Central Salt Administration Accounts; District Reports.

of bigger receipts through increased consumption. Conversely, his policy of deliberately expanding sales was admirably geared to the new needs of Manchuria for salt.

Ch'ang-lu. Except in the matter of free trade, Ch'ang-lu was the most successful of Dane's salt divisions, closely approximating his blueprint and showing gratifying increases both in salt releases and in revenue.

In the last year of Hsüan-t'ung, Ch'ang-lu produced 5,236,800 taels of revenue (= $7,855,200); in 1914, as Dane said, "without any large increase of taxation," the revenue stood at $12,650,260, and by 1922 it was $15,444,453.[30] In 1911, Ch'ang-lu ranked third in terms of salt handled and fourth in terms of revenue, but by 1919, it had risen to second place on both counts, exceeding Huai-nan for salt and Szechwan for revenue.

Ch'ang-lu was well suited to Dane's system. While salt could in theory be produced almost anywhere along the Chihli coastline, in fact, thanks probably to communication advantages, production was largely concentrated at the Feng-ts'ai and Lu-t'ai works within a comparatively short radius of Tientsin. The three huge depots Dane had constructed at Han-ku, T'ang-ku, and Teng-ku could therefore easily control the salt at its source. The only difficulties, the outlying salt works in Yung-p'ing prefecture in the northeast of the province and the small Yen-ch'eng-Hai-feng group near the Shantung border, were met in the first case by the establishment of the Shih-pei assistant inspectorate in 1914 and in the second by the closure of the works in 1916.

Ch'ang-lu too, centrally placed in relation to China's new railway network, was able to exploit the opportunities presented by the less rigid view of *yin-ti* frontiers held by the foreign chief inspector. Thus, Ch'ang-lu bittern block (magnesium chloride, a valuable by-product of salt used in bean curd and tobacco manufacture) competed with that of Fengtien along the Peking–Mukden line, while in northern Shansi the Chin-pei collectors hoped that Ch'ang-lu salt brought in along the Chengting–Taiyuan railway might in time eliminate the virtually untaxable local earth salt. In Shantung, the salt merchants, annoyed by the high prices charged by their local producers, forced a reduction by threatening to import cheap Ch'ang-lu salt along the Tientsin–Pukow line. The Ho-tung salt area was invaded by both the Peking–Hankow and Lunghai lines, while the former railway also carried Ch'ang-lu salt into Huai-pei and even into the Szechwan districts of Hupei. In 1918, the Huai-pei inspectors had to admit that in southwest Honan, "The issue for the Jukwang district was absolutely nil, Huaipei salt being entirely driven out of the market by Changlu salt, which has the advantage of rail communication." [31]

With this record of success behind it, the failure in Ch'ang-lu of free trade was the more disappointing, and it has already been noted that the defeat here led Dane to abandon all hope of making his system universal. At first sight, it seems strange that Dane could enforce free trade in distant Szechwan and Kwangtung, yet could not do so in Ch'ang-lu. The ability of the Ch'ang-lu merchants to resist the chief inspector rested on their traditional alliance with the central government, which went back to the time of the K'ang-hsi Emperor and had been noted by the nineteenth-century salt expert, Wang Shao-chi: in times of emergency, the merchants supplied funds to the government in return for privileges and protection.

This alliance still operated in 1915. Dane noted that the monopoly company which effectively nullified his free trade plans in

1915 "included a number of very influential names" and referred to the "rumoured intention of the Minister of Finance to levy a contribution of $700,000 from the merchants" in return for a confirmation of their privileges.[32] Similar connections between officials and salt merchants existed of course in other parts of the country; the advantage of the Ch'ang-lu merchants was that their relations were specifically with the central government. This was not only a matter of geography: Yüan Shih-k'ai had built up his power as Chihli governor-general, and Chou Hsüeh-hsi, the minister of finance, was a former Ch'ang-lu salt controller. In Szechwan and Kwangtung, the central government by abolishing monopoly was giving up other people's money, namely that of the provincial authorities; in Ch'ang-lu, Dane was asking it to give up its own. Ch'ang-lu, therefore, with the doubling of its revenue, its increase of over a million piculs in taxed salt, and its disappointing failure in free trade, accurately mirrored the strength and weakness of Dane's type of reform.

Kwangtung. Unlike Ch'ang-lu, the revenue record in Kwangtung was unimpressive, except in the new assistant inspectorates at Swatow and Pakhoi. In 1911, the entire salt revenue of the division had been farmed out to a merchant consortium for 5,800,000 taels (= $8,700,000): in 1922, after ten years of inspectorate administration, it had increased only to $10,530,412. Unlike Ch'ang-lu again, the sales area of Kwangtung salt did not expand: the promising market in Hunan, which in 1911 had led to a demand for redrawing the divisional boundaries, was largely lost with the revival of Huai salt, while the salt division suffered generally from the failure to complete the Hankow–Canton railway.[33] On the other hand, in growth rates for taxed salt and in the liberalization of the trade, Kwangtung did comparatively well.

In 1913, the area possessed three long-distance water routes: the West River route via Wuchow to Kwangsi, the North River route to Hunan, and the Han River route via Swatow to southern Kiangsi and western Fukien. The first two presented few difficulties after the collapse of the tax-farming system during the revolution, and they could be adequately controlled through the existing machinery at Canton. The Ch'ao-ch'iao region of Swatow, however, was again farmed out after the revolution, thus depriving the inspectorate of effective control, and it was not until 1916 that the contract ran out and the tax-farming system was ended. Besides these major routes, the inspectorate was also confronted with the problems of the land route from Pakhoi to Kwangsi, and of the short-distance water routes serving the thickly populated districts on either side of the Canton estuary. The first was dealt with by the establishment of the Pakhoi assistant inspectorate, the second by the elimination of tax farming and the setting up of regular collectorates. Thanks to these measures, salt taxed in Kwangtung rose from a prereform figure of 1,954,821 piculs to 4,516,809 piculs in 1917, the last year undisturbed by serious political disorder.[34]

In 1911, free trade had been established in the central, northern, and western *kuei* (lit., "farms") or monopolies: that is, in the Canton area, in the Shiuchow and Hunan trade, and in that to Kwangsi. In September 1914, the salt commissioner Ou Lien, himself a salt merchant, attempted to reintroduce the monopolies, offering a substantial bribe to Peking for permission to do so. When taking out their licenses, the privileged merchants had to deposit security with the commissioner, either in cash or in bonds of the domestic loan. In the latter case, less deposit was required. Thus, as Dane put it: "The inducement to the Central Government to approve the scheme was the amount, which the Commissioner proposed to subscribe to the Domestic Loan." [35]

Dane, however, was able to counterattack. In July 1915, as part of a package deal whereby in return for the release of $7.5 million of the loan funds originally assigned for salt reform the Chinese government agreed to accept Dane's views on a number of issues, the cancellation of the Kwangtung scheme was ratified and free trade was confirmed in the three *kuei*. The following year, monopoly, along with the tax-farming system, was eliminated from Swatow, Pakhoi, and the Canton estuary, so that the whole salt division now formed a free trade area comparable to that of the Szechwan *yin* district. In Kwangtung, therefore, unlike Ch'ang-lu, Dane was able to outbid the monopolists and get his system accepted.

CHAPTER 5

The Eastern Quadrants

To the east of a Peking–Canton line lay the salt divisions of Shantung, Huai-pei, Liang-che, and Fukien, all regions where, in the long run, the inspectorate was successful. Dane laid down the basic policies, though in many cases they did not bear fruit till after he had left China.

Shantung. The problems of control in Shantung were not dissimilar to those in Fengtien: a long coastline with scattered, inaccessible salt works as well as the existence of leased territories, Weihaiwei and Kiaochow, whose exemption from Chinese jurisdiction made them ideal bases for smuggling. These difficulties were compounded by the disruption caused by Japanese military operations in 1914 and by a local increase in tax in 1915.[36] The Japanese, moreover, proved less cooperative, or less efficient, masters of the concession than the Germans: "Salt produced in the leased territory of Tsingtao was formerly not permitted to be exported to the interior of Shantung, but in 1915 large quantities of such salt were smuggled by rail into the districts along the Tsingtao–Tsinanfu Railway."[37]

These problems were eventually overcome. In the first place, control of salt using the main trade route, the Hsiao-ch'ing River, which had formerly commenced only at Tsinan, the railhead for the Tientsin–Pukow line, was now pushed back to the works by the establishment of an assistant inspectorate at Yang-chia-k'ou near the mouth of the Yellow River.

The crucial difficulty, however, lay in the mountainous, promontory area of east Shantung. Here between 3 and 4 million piculs of salt were produced annually and exported, without paying tax, either to the rest of Shantung or overseas to Korea, the Yangtze provinces, and Hong Kong, playing havoc with the sales of local tax-paid salt. The problem, which Dane admitted he had at first underestimated, was not tackled seriously until 1918, when C. G. G. Pearson, then district inspector, drew up a plan for controlling the salt works by establishing a line of collectorates and suboffices along the coast from Chefoo to Kiaochow. This plan, put into effect the following year, met with strenuous local opposition, especially along the south coast of the promontory. The inspector reported: "On June 8th and 9th, clearly by pre-arrangement . . . armed bands methodically attacked and destroyed the stations in Lai Yang and Hai Yang districts, fourteen members of the Preventive and Inspectorate services, and seventeen villagers associated with the Department, being brutally killed." [38] Similar attacks took place in Chi-mo district to the south, while in Shih-tao to the north, where more violence was feared, tax collections had to be abandoned. Despite the setback, which was the worst outbreak of violence the inspectorate had to face, control over promontory salt was gradually achieved: by 1922, 1,482,100 piculs were paying tax.[39]

Pearson also took steps to deal with the foreign concessions. In 1918 he negotiated an antismuggling agreement with the British commissioner at Weihaiwei, and his deputy, T. Koizumi, assistant inspector at Chefoo, made a similar agreement with the Japanese authorities. It made a number of concessions to the Japanese over the use of the railway, but the inspector subsequently claimed that "since this Agreement was concluded illicit transportation has been reduced to vanishing point." The problem of Kiaochow

Table 17. Salt releases in Shantung, Huai-pei, Liang-che, and Fukien, 1913–1925 (piculs).

Year	Shantung	Huai-pei	Liang-che	Fukien
Prereform	2,095,744	1,440,000	1,700,620	772,000
1914	2,582,037	–	1,904,278	2,194,197
1915	1,219,803	1,462,406	2,072,893	1,774,302
1916	1,069,614	2,952,759	2,308,789	861,799
1917	1,261,207	1,770,069	2,579,215	1,292,807
1918	1,298,737	2,116,198	2,758,492	705,737
1919	1,869,510	4,026,830	3,074,616	754,126
1920	1,992,900	2,828,563	3,134,457	1,411,402
1921	2,441,268	2,472,540	3,248,335	1,521,044
1922	3,521,038	2,759,484	3,222,918	1,537,003
1923	4,886,985	2,327,510	3,350,726	1,414,481
1924	4,376,555	1,799,877	3,299,106	821,811
1925	3,792,831	2,768,902	3,289,607	714,757

Source: Chang Chien, *A Plan for the Reform of the National Salt Administration;* Central Salt Administration Accounts. For the period following the reform, the figures for Liang-che and Sung-kiang inspectorates are combined in order to cover the same area as the prereform entry for the undivided division.

received a definitive solution with the rendition of the concession on December 10, 1922.[40]

The figures for taxed salt in Shantung (Table 17) reflect a decline in the first years of the Japanese occupation and a recovery of control from 1917 onwards. The maintenance of a steady increase in taxed salt from 1917 to 1922 can be related to the general economic development of the area by the Japanese: by 1922, for example, the Japanese military government had spent 126 million yen on reconstruction,[41] in anticipation of their own permanent occupation.

Huai-pei. In contrast with Shantung, Huai-pei was an easy salt division to reform along inspectorate lines. Its salt came from a short strip of seacoast in northern Kiangsu, and the trade followed well-defined water routes, either west to Anhwei or south to the Yangtze ports.

Effective control over both routes was assured by the establishment on April 1, 1914, of the separate inspectorate of Huai-pei, located at Pan-p'u on the north Kiangsu coast, which thus formally separated the region from Huai-nan. Taxes were consolidated into a single collection at Pan-p'u, and the additional collections at Hsi-pa and Cheng-yang-kuan were abolished.

Trade was liberalized: for example, the Cheng-yang-kuan government transportation office for northern Anhwei was dissolved in March 1914. In December of the same year, the privileges of the *p'iao-shang,* the merchants who handled the trade as far as the Grand Canal, were canceled, and Dane was able to defeat an attempt to revive them in 1915 by the same package deal (effected through his control of the loan funds) which saved free trade in Kwangtung. Beyond the Grand Canal, restrictions on markets and suppliers continued down to 1921, but they were finally extinguished against compensation. Structurally, therefore, Huai-pei realized the ideal.[42]

The figures for taxed salt (Table 17) reflect this success and show an increase of nearly 100 percent during the first decade. The figures, when broken down, show development in all areas but one. Before the reform, Huai-pei taxed salt totaled 1,410,302 piculs, of which 177,817 were consumed in the eleven Kiangsu districts of the division, 1,034,485 were sold in northern Anhwei, and a further 198,000 went to southeast Honan. By 1920, the picture had changed considerably: local consumption districts, now including a segment of Shantung formerly untaxed, accounted for 518,300 piculs; northern Anhwei took 1,711,150 piculs; but Honan took no Huai-pei salt at all, the market having been entirely lost to Ch'ang-lu. Huai-pei, however, was more than compensated by an export of 688,361 piculs from Pan-p'u to the Yangtze ports of Huai-nan, a result of Dane's more liberal view of *yin-ti* boundaries.[43]

In addition to this direct shipment, there was also an indirect export to Huai-nan from Huai-pei via Yangchow. This was not shown in the Huai-pei release figures since it paid tax at Yangchow and not at Pan-p'u, but it formed a significant part of the Huai-pei trade. In 1913, the reports give a quota for this indirect shipment of 1,119,000 piculs; in 1920, the quantity so transported was 1,723,164 piculs. This figure was in fact exceptionally low: the preceding and following years both show an indirect import from Huai-pei of over 3.5 million piculs. In 1919, the peak year for Huai-pei since there was an unrepeated export to Japan of 919,600 piculs, total salt handled by the division reached 7,630,000 piculs, three times the aggregate prereform quotas.[44]

Liang-che. In Liang-che, the inspectorate was confronted with the most difficult of the salt divisions from an administrative point of view. It exemplified the problem, never successfully tackled by the old regime, of a heavily urbanized region lying close to a widely dispersed group of salt works.

In 1913, Liang-che included Chekiang, Kiangsu south of the Yangtze, southern Anhwei, and parts of Kiangsi. Salt was produced from Sung-kiang to Wenchow, but the old regime, true to its principle of paying most attention to long-distance waterborne salt, tended to concentrate on the salt traveling inland by river along the tea route from Hangchow to Hui-chou and southern Anhwei. The main centers of population, however, were along the seacoast, and here, because the salt works were so close to the heavily populated urban areas, effective taxation was simply not attempted. In the Shanghai Settlements and at Ningpo and T'ai-chou, an inefficient tax-farming system was practiced, and, as a result, out of a total trade for the division of 3,500,000 piculs, only 1,700,620 piculs had been brought to taxation in 1911.[45]

The most acute problem was that of the Shanghai area, the so-called Su Wu-shu districts of Kiangsu south of the Yangtze, which received their salt from the Yü-yao and T'ai-shan works on the Chekiang side of Hangchow bay. On a short-distance water route such as this, the old gabelle, with its headquarters at Hangchow, found it difficult to isolate the salt junks in the rapid turnover of local shipping. Dane dealt with this difficulty by establishing a new inspectorate at Shanghai on June 1, 1914. The new inspectorate, that of Sung-kiang district, was headed by an able Russian, G. Brauns, who brought the trade under effective supervision as soon as it entered the Whangpu River. Subsequently a new assistant inspectorate was established at Yeh-hsieh to deal with the outlying Sung-kiang works.[46]

Brauns also revised the terms of the Shanghai Settlements tax farm. In 1913, the tax farmer, a Mr. Li, paid the government $96,000 annually and in return received the right to import into the Shanghai Settlements, free of all further duty, 96,000 piculs of salt. This was supposed to be a maximum quantity, but as the inspectors noted, "Control over the quantities of salt actually imported into the Settlements by the Farmer can hardly have been effective, as the staff of the Woosung Checking Office was paid by the Farmer and not by the Administration." [47] In 1917, under the revised arrangements, the new farmer, Ku Chao-te, was to pay $1.35 a picul on an import of 116,016 piculs of salt, giving an annual revenue of $156,621. The Woosung checking office was now placed under inspectorate staff, and in 1919 the restriction on the quantity which could be imported was lifted. In that year, when the rate had been raised to $1.50 a picul, the Shanghai farmer in fact imported 144,247 piculs, giving a revenue to the government of $216,370.[48] The value of the farm had been more than doubled.

The free trade policy was less vigorously pursued; monopolies continued in the Su Wu-shu, in the Shanghai Settlements, and along the Hangchow-Hui-chou route; quota limitations, however, were lifted in the Su Wu-shu and Shanghai, and merchant monopoly was eliminated at Shao-hsing and T'ai-chou.[49] The increased figures for taxed salt in Liang-che thus primarily represent increased control, though in Shanghai the general prosperity of the city from 1912 to 1921 (its trade was increasing faster than that of any other major center except Dairen)[50] was also a factor. These two factors together ensured that salt subjected to tax doubled in the Liang-che area in the first decade. This was a considerable achievement in this complex, difficult region.

Fukien. Serious reform in Fukien began only in 1918 following a visit by Dane in the winter of 1916–17, and after this it was not long before the region became a battlefield in the wars between north and south. Nevertheless, the improvements produced were not negligible.

Southern Fukien, with salines extending from Foochow to the border of Kwangtung, was well placed for salt production: "The influence of the Yangtze river has ceased to affect the density of the sea brine, and the climate appears to be drier than the climate of Chekiang or Kwangtung." [51] Production, however, was liable to severe fluctuations due to damage in the wake of typhoons and to the vagaries of outside markets, one of the features of the Fukien trade being the relative importance of the export trade to other salt divisions. In an average year, 750,000 piculs were exported to the Swatow region of Kwangtung and 250,000 piculs to southern Chekiang, particularly Wenchow.[52] Fukien, indeed, with its large but erratic production, acted as the reserve supplier for the whole south China coast.

Before the establishment of the foreign gabelle, the weight of salt administration fell principally not on the export but on the domestic trade, up the Min River from Foochow. After the reforms of 1912, this route was reasonably well controlled by the salt commissioner and his staff. Dane, therefore, concentrated initially on salt exported from Fukien to other regions.

He found that likin stations had already been established at the main export points, Lien-ho north of Amoy and Chao-p'u near the Kwangtung border, but the revenue was in effect farmed out to the officials in charge: "Each Collectorate was assessed at a certain amount payable yearly to the Commissioner and its liabilities were regarded as discharged as soon as the required amount had been paid." [53] This system was ended: from 1914 all taxes collected at the likin stations were paid into the Bank of China at Foochow, and from 1917 all the staff concerned became subordinates of the inspectorate. The establishment of a collectorate at Wenchow in Chekiang and of an assistant inspectorate at Swatow in Kwangtung, the two chief ports of entry for Fukien salt to the neighboring provinces, also served to strengthen control: the export trade was in fact now better controlled than the domestic trade.[54]

As regards trade liberalization too, Dane at first made few changes in the internal arrangements of Fukien. In 1912, the local Chinese salt reformers, led by Liu Hung-shou, had set up a completely government-operated monopoly of transportation and sale within Fukien. This kind of reform was at variance with Dane's system, and in most cases he abolished the government transportation offices. In Fukien, however, he decided otherwise: "it was considered undesirable to take any action which might hamper the Commissioner in his management of the monopoly, and thereby detract from the value of the experiment in administration which the monopoly afforded." [55] Chinese-style reform should in this instance be given its head.

However, Fukien too was eventually brought into line. In 1917, after a visit to the area, Dane reversed his policy of acquiescence in government transportation. The reasons for this will be examined in more detail in Chapter 6 when Dane's relations with the Chinese salt reform movement are considered, but basically he felt that the "experiment in administration" [56] had been given a fair trial and had failed. From 1918 onwards, the official monopoly was gradually dismantled, until in 1921, with the removal of restrictions in the Amoy region, the inspectors could write: "Thus free trade was established in all the areas under the control of the Fukien Administration." [57]

The Western Quadrants

In the west, there were two regular salt divisions, Ho-tung and Yunnan. In addition, there was a third region in the northwest, stretching from Jehol to the Koko-nor, managed by various territorial agencies but unified by being supplied for the most part by trans-border salt from Tibet or Mongolia. In none of these areas were external conditions at all propitious for the inspectorate (Table 18).

Ho-tung. Comprising southern Shansi, western Honan, and eastern Shensi, Ho-tung seemed well suited to Dane's reform, since virtually all the salt came from a single source, the An-i-hsien salt

Table 18. Salt releases in Ho-tung, the northwest, and Yunnan, 1913–1925 (piculs).

Year	Ho-tung	Northwest	Yunnan
Prereform	1,589,400	–	512,300
1914	1,110,978	–	509,370
1915	1,854,000	468,542	806,417
1916	831,000	661,383	834,998
1917	1,469,400	561,558	761,137
1918	1,470,600	695,290	820,375
1919	1,269,300	691,028	805,796
1920	1,082,700	611,339	766,120
1921	1,192,800	893,398	776,757
1922	1,185,900	884,033	744,790
1923	1,156,000	–	690,976
1924	1,137,000	–	628,581
1925	1,443,000	1,045,236	632,192

Source: Chang Chien, *A Plan for the Reform of the National Salt Administration;* Central Salt Administration Accounts. The northwest figures were formed by adding the entries for the collectorates of K'ou-pei, Chin-pei, and Hua-ting in the Central Salt Administration Accounts.

lake, around which there was a protecting wall dating from Sung times. As Dane wrote: "The reform of the Hotung administration . . . was simplicity itself. All that was necessary was to impose a single direct duty and to collect it at one Collecting Office at the Lake, and to see that no salt was transported from the Lake without payment of this direct duty." [58] Some reforms along the usual lines were indeed accomplished: government transportation in the thirty-three districts in which it applied was eliminated, as were the second collections of tax in Honan and Shensi; outlying salt works were suppressed; and a trench was dug inside the crumbling wall to reinforce inspectoral control of salt leaving the lake.[59]

Despite these reforms, Ho-tung in the period 1914 to 1925 only once surpassed its prereform figure for salt released. The region suffered from political disorder: in 1916, the inspectors wrote, "From May to August we were never sure of being unmolested by bandits"; and the following year, "The transport trade into Shensi was entirely stopped for over three months." [60] Natural calamities aggravated the situation: in 1919–20 and again in 1922, there was severe drought and famine, with the inspectors reporting in the latter year that "The whole marsh within the lake precincts is now absolutely dry." [61] Not only was there a shortage of brine, but higher prices for the transportation animals' fodder and generally reduced economic activity affected the salt trade adversely.

Finally, Ho-tung was the victim of Ch'ang-lu's dynamism: as was noted earlier, it was losing its markets in Honan to sea salt brought in by rail. In 1915, 538,242 piculs passed the checking stations en route from the salt lake to the cities of western Honan, Lo-yang, Ju-chou, and Nan-yang; in 1920, only 195,660 piculs passed. The inspectors commented on this decline: "It has caused many of the merchants to close down their business. Only 41 merchants were selling their salt in Honan to retailers against 58 in 1919 and 77 in 1918." Politics, drought, and railways combined to make the area's trade decline, and Dane's reforms in Ho-tung found themselves involved in the contraction.[62]

The northwest region. Beyond Ho-tung lay the northwest. In this area too, external conditions impeded reform: the size of the area, the number of petty salt works, and the endemic political disorder combined to present a formidable obstacle to success. Salt consumed in the northwest came from the salt fields of Fengtien and Ch'ang-lu, from the earth salt deposits of northern Shansi, and from the large number of salt lakes in the Tibetan–Mongol borderlands. Before the reorganization, official control over this widely spread network of routes was very imperfect. In the Hua-ting salt administration (so-called from the salt lakes at Hua-ma and Ting-pien, not far from Ninghsia), which comprised Kansu and western Shensi, the right of control was farmed out to three merchant companies for an annual payment of $560,000 which was often in arrears. Further east, the Chin-pei (north Shansi), K'ou-pei (Kalgan), and Jehol transportation offices concentrated on the main trade routes only.[63]

Dane was more ambitious. In Hua-ting, tax farming was abolished and a line of collectorates was set up from the Koko-nor to the Ordos to control not only the Hua-ma and Ting-pien lakes, but also Tibetan salt and salt from the lesser lakes of the Alashan region, Abalai and Alabruka.[64] In Chin-pei, effective control was for the first time attempted over the sale of earth salt, which supplied the greater part of northern Shansi. Here, "The chief difficulty . . . is the number and scattered location of the factories."

This was met by a policy of forcible closure: "It was therefore proposed to discourage the intermittent production . . . and to make it a condition of license in Chunglu and Nanlu that tax on not less than 48 piculs per annum be paid." [65]

These measures produced considerable improvement in the figures for salt taxed in the northwest, but the administrative costs were high, and it was doubtful if the revenue received repaid the trouble of collecting it. In Hua-ting collectorate in 1916, the tax-farming system brought in $446,080; in 1922, there was a gross tax collection of $1,012,032, but administrative expenses of $569,970 reduced the net revenue to $442,062. In Chin-pei too, expenses in that year came to over a third of the gross revenue.[66]

Trade in the northwest was liberalized. The Jehol and K'ou-pei transportation offices were abolished in 1914, free trade accompanied the elimination of tax farming in Hua-ting in 1918, and in 1921 government transportation of Mongolian red salt ceased in northern Shansi.[67] Whatever impulse, however, these measures may have given trade in the area was largely nullified by adverse economic and political circumstances. In Hua-ting, transportation was a recurring problem. Sometimes, the camel drivers were reluctant to take the salt to a given destination in case there was no return freight to be had; at other times the drivers would not come to the salt lakes for fear of losing their camels to local condottieri. Tibetan salt might be held up by an outbreak of yak sickness in the Koko-nor.[68] Similar conditions were found elsewhere: for example, in 1920 in K'ou-pei collectorate, which covered Jehol and Chahar, all salt carts at the Ahlatangkatasu lake were commandeered by troops going to Urga to suppress the Mad Baron's revolt.[69]

Yunnan. In Yunnan, the figures for salt released showed a definite advance. Nevertheless, here too, despite the energetic measures of successive French inspectors, the external conditions of the province — its size, its primitive economy, the prevalence of malaria, and the activities of the warlord T'ang Chi-yao — frustrated the reform.

As in Ho-tung, the problem of making the salt tax comprehensive was relatively simple. There were only three groups of wells: Hei-ching near Kunming, Pei-ching in the Tali region, and Mo-hei-ching in the south not far from Szemao. Moreover, the mountainous terrain, by limiting the number of possible salt routes, facilitated control. Since the beginning of the nineteenth century, the taxation-at-source system (*chiu-ch'ang cheng-shui*) had been theoretically in force, and Dane praised the old regime in Yunnan as "the most simple and practical of all the systems of administration in force in China." [70]

In fact, as he discovered, "A system of farming the revenue to the Officers in charge of each of the three groups of wells appears to have been generally practised," [71] and to get effective control assistant inspectorates had to be established at Mo-hei-ching and Pei-ching and eventually at Hei-ching as well. As so often, it was only intensive control exercised by the inspectorate which made the system operational.

The inspectorate adopted a dual policy towards the commercial organization of the Yunnan salt. On the one hand, marketing was liberalized: government transportation was abolished, as were the separate consumption areas for the three groups of wells.[72] On the other hand, the inspectorate took a more active interest in the production of salt in Yunnan than in other divisions: the use of coal in place of wood in boiling was encouraged, a system of accident compensation was started, and medical facilities at the works were provided. In 1921, the inspectors went so far as to declare that "Mines should be worked entirely under Government control and repairs to same paid from the Salt Revenue, while miners and other workmen should be under the full control of the Salt Administration, by which they should be paid." [73]

This incipient *étatisme* was due less to any settled policy of the gabelle — Dane himself was opposed to the government becoming involved in salt production — than to the desire of the inspectors to get trade going in Yunnan despite the discouraging economic circumstances of the province. By 1922, political disorder and the collapse of the wartime tin boom were choking off whatever expansion there had been. The inspectors commented: "Business in Yunnan is practically stagnant. The principal article of export, namely tin, has been much neglected during 1922, owing to the decreased prices ruling on the European markets . . . Brigandage has been rampant throughout the year and has greatly hampered business transactions." [74]

In every direction, Dane's reform was conditioned by the relation in which the Chinese environment stood to him and he to it: the extent to which he forwarded or frustrated its needs and aspirations, particularly those of the several million people whom Ching Pen-po estimated were vitally concerned with salt administration. This relationship, decisive for the final success of modernization, will form the subject of the next chapter.

6 | Dane and the Salt Interest

In early republican China, the various groups connected with the salt trade formed a powerful vested interest, comparable to the opium interest of nineteenth-century Canton, the winegrowers' pressure group of Napoleonic France, or the West Indian sugar interest of eighteenth-century England.[1] China was not alone in having an articulate salt interest: in other traditional societies as well, salt, "l'objet premier du grand commerce pondéreux," [2] had given rise to diplomatic, fiscal, and commercial pressure groups, similar to those identifiable in China. What distinguished the Chinese salt interest was its conspicuous and comprehensive organization; and a comparison with another premodern salt interest, that of sixteenth century Europe, shows that these special characteristics were the result chiefly of the topography and distribution of the trade. Admittedly, the Chinese empire was a political unit, while Europe was not, but the governors-general enjoyed in practice a considerable measure of autonomy, while in Europe, on the other hand, the Habsburg predominance drew together much of continental Europe under the Spanish aegis. Fundamentally, it was geography, not politics, which determined the character of the salt trade.

First, in China, the principal salt routes were riverine and this made possible control of the trade by a series of checkpoints, which could then combine to form a single system, covering a part, or the whole, of the route from producer to consumer. In the Yangtze valley, for example, checkpoints extended all the way from the salines to the sales suboffices in the up-river provinces.

The salt junks en route from Shih-erh-wei to Hankow had their papers checked and cargo examined at Nanking, Ta-t'ung, and Wu-hsüeh, as well as having to report at Wuhu and Kiukiang. Between resale at Hankow and finally passing out of official control at one of the suboffices, the salt would pass through at least four further checkpoints. All these were combined under one authority, since both the sales headoffice at Hankow and the twenty-two sub-offices scattered over Hupei were responsible to the Liang-kiang governor-general as head of the Liang-huai salt division rather than to the local Hu-kuang authorities.[3] The governor-general and his salt administration collected the taxes of the whole area and enforced the monopoly of purchase and sale granted to the merchants. The degree of control achieved may be illustrated from the elaborate and extensive salt codes, in which even such minor items as the transit times, from points A to B, were prescribed in officious detail.[4] This high level of organization was made possible to a large extent by the geographical factor of riverine traffic.

In Europe, on the other hand, the principal routes were maritime: from Cyprus, Corfu, and Sicily to Venice; from southern Spain and Sicily to Genoa; from Portugal and the French Biscayan ports to the Low Countries and the Baltic.[5] Close control by toll stations or checkpoints was impracticable, except at the Baltic Sound. On these maritime routes, no system of control, complete or partial, was feasible, and monopolies were correspondingly precluded. In the seventeenth century, for example, Setubal, "la capitale mondiale du sel," was the principal supplier of salt for northern Europe, and its chief customer was Holland, "premier consommateur du monde." [6] Neither, however, could exert a monopoly over the other: Setubal sold to the English, the Danes, the Swedes, the Hansards, and the French, as well as the Dutch; while the Dutch obtained their salt not only from Setubal, but also from the French Biscayan ports, from Scotland, from Araya on the Venezuelan coast, and from the Cape Verde islands. Even strictly local monopolies were incomplete: among the Dutch salt merchants there were no monopolies, and the same was true of the English; while in Portugal, the *casa da contratacao do sal* claimed a monopoly of purchase for the domestic trade only, not for the export. Only on routes like the Loire and the Rhone, where conditions approximated those of China, was monopoly feasible.[7]

The unified control of the trade possible on riverine routes might, in theory, have been exercised from either the producer's or the consumer's end of the business. In China, topography favored the producer, for the coastline, which, as in Europe, was the principal source of salt, was short in relation to the hinterland it served, and this had the effect of increasing the power of producers over consumers because there were few convenient sources of supply. Kweichow, for example, which Hosie described as the Switzerland of China, had less choice of salt supply than its European counterpart. In normal circumstances, all but one of the province's fourteen prefectures were supplied with Szechwan salt. The remaining prefecture, Li-p'ing, was supplied from Canton, while limited quantities of salt entered Kweichow from Yunnan. Even in emergencies, Canton and Yunnan could not meet the whole demand from Kweichow, the first because of transportation difficulties across the Yangtze–West River divide, and the second because its total production came only to half of Kweichow's requirements. In 1914, for example, prohibitive taxation on the Kweichow border temporarily cut off supplies from Szechwan, and this produced famine prices of 300 cash a catty for salt: neither Canton nor Yunnan could make up the deficit. The absence of alternative suppliers inevitably made Szechwan the dominant partner in the exchange.[8]

Switzerland, on the other hand, was normally served not only by a variety of inland sources — Savoy, Franche Comté, Lorraine, and the Tyrol — but also by two major maritime suppliers: by Languedoc salt coming via the Rhone and Lake Geneva, and by Venetian salt arriving via the Po, the Italian lakes, and the Simplon. In the eighteenth century, for example, when the Savoyards occupied Domodossola and cut off the supply of Venetian salt, the Valais was able to play off French salt against Savoyard salt.[9]

In contrast with Europe, then, the advantage in China lay with the producer, and the gabelle, in both its administrative and commercial aspects, was organized from the producer's end of the business. This method of organization affected the public standing of the industry. A business takes its popular image from its head-office, and the local influence and importance of salt officials and merchants in the producer areas had the effect of making the trade generally conspicuous. On home ground, the salt trade was preeminent: its trading centers were salt cities, and the officials and merchants of the trade were ex officio leading citizens. Ho Ping-ti has shown how the salt merchants dominated the social life of Yangchow,[10] while at Tientsin Detring was sufficiently impressed to note that the salt merchants were "the wealthiest people in the province." [11] The senior salt officials ranked next only to the provincial treasurer and judge, above taotais, prefects, and provincial censors. This prominence lent the whole interest a cachet of which few Chinese were unaware. The leading role of the salt interest in the producing areas may be contrasted with the minor role played by salt importers in the consuming centers: in Peking, the gabelle merited only 3 representatives in the chamber of commerce out of a total of 270 members and had no representatives at all among the 35 members of the board of directors.

In Europe, on the other hand, initiative rested with the consumer and the weight of the organization fell at the consumer's end of the business. In Portugal, Frédéric Mauro states, "les transporteurs de sel sont les Hollandais";[12] the Genoese farmed the Spanish salines; the Venetians went to Cyprus; and Aksel E. Christensen shows the leading Delft firm of Claes A. van Adrichem organizing the import of salt into Holland from the French Biscayan port of Brouage and from San Lucar de Barrameda at the mouth of the Guadalquivir. As a result, the salt trade was bound to appear insignificant in the total volume of trade at Amsterdam, Genoa, Venice, and Delft. An importer of salt at these centers acquired no particular influence or importance through his association with the trade: salt was just another consumer commodity. At Delft, for example, van Adrichem served as burgomaster and was one of the half-dozen richest citizens, but he "first and foremost was a corn-merchant." [13] In fifteenth-century Genoa, the Centurioni imported salt from Spain, but they were primarily merchant bankers whose interests embraced a wide variety of commodities: bullion, mercury, and grain. In Scotland, William Kinnereis, town councilor and eventually provost of Dundee, traded in salt, but his main business was the import of wine: in 1618, his firm imported thirteen cargoes from Bordeaux, "four of them of salt, the other nine predominantly of wine." [14] A city may specialize in exporting salt; no city specializes in importing it, unless it is primarily a fishing center, in which case salt is an industrial raw material rather than a foodstuff. Salt in Europe, therefore, remained anonymous and inconspicuous: it was ballast for returning ships, "when all better employment fails," as an English state paper for the reign of Charles I put it.[15]

Europeans, therefore, when they came to China, were unprepared for the comprehensive and prominent trade complex that had grown up around the salt trade, and tended to take a narrower view than the Chinese of the size and scope of the industry. Sir Robert Bredon's question to the customs commissioners, "How many people are engaged or employed in the salt business — officials, workers and traders?" prompted an equally narrow answer. H. E. Wolf of the Anhwei salt likin collectorate replied for his district: 969 officials, 2,629 workers, and 1,638 traders. He was, however, uneasily aware that a wider answer would have been appropriate: "This office has no means to ascertain, or even calculate with some degree of accuracy, the number of men engaged in, and living on, the salt trade. Their number is legion." P. C. Hansson too, in the Liang-huai, felt obliged to extend the question's frame of reference: of the 369,000 people, 230,000 were workers at the *ch'ang,* 24,090 were official personnel, and 115,000 were engaged in commerce, of whom 80,000 were the crews of the 2,000 salt junks.[16]

Chinese estimates, more sensitive to all the affiliations of the trade, were much higher. Chang Chien reckoned the total of those involved in the salt business at several million, while the Szechwan salt merchants, in a petition to their provincial assembly in 1913, stated that a majority of the province's population was involved, directly or indirectly, in the salt trade, and that several tens of millions would be out of work if the wells had to close.[17] Given the multiple linkages of the salt industry, its demands on boatmen, rice peddlers, ironmongers, and its supplies to fishermen, vegetable salters, soy sauce manufacturers, pastoralists, figures of this order are not implausible.

It is not surprising that so extensive, prominent, and highly organized a commercial interest was capable under pressure of engendering articulate subgroups from among those either threatened by change or desiring to exploit it. In the early republican period, three sections in particular were provoked by Dane's reforms into expressing their views: the officials, Dane's colleagues in the administration; the merchants, who were paying the tax and conducting the trade; and the reformers, who set the tone of public discussion. Predictably, they neither agreed in their original appreciation of the situation, nor concurred in their verdict on Dane's performance; but, to generalize their positions, it is broadly true that while they all disagreed in theory with Dane's ideas, in practice the more progressive elements of each lobby were won over to cooperation as Dane's program of reform took shape.

In the first place, on the level of theory, merchants, officials, and reformers shared a belief in the necessity of economic controls, and a disbelief in the optimal working of market forces.

Joseph Needham has stressed the importance in Chinese natural science of the idea of immanent, self-regulating order, a concept which would translate in socioeconomic terms as a theory of laisser faire. Chinese salt experts, however, seem to have felt that, on the contrary, nature if left to itself would tend to imbalance and economic dislocation. Ching Pen-po in an editorial in the fourteenth number of the *Yen-cheng tsa-chih* (Salt administration magazine; hereafter *YCTC*) described the chaos which would develop if free trade were introduced: "no group of men will be designated as transportation merchants, and there will be no fixed quota for transporting salt. As a result, there will be no limitation on the amount being produced, and planned distribution will be impossible. Then there will be famine here and glut there, or else violent fluctuations in price." [18] Yao Yü, Kwangtung salt commissioner, in October 1915 expressed similar doubts: "The farming system has indeed led to many abuses, but under the system of free trade, it is difficult to fix the responsibility on anyone . . . Unless there are some restrictions, matters will surely get beyond control." [19]

At the social level, the Chinese argued, order must be a social creation, achieved by deliberate intervention. A writer in the *T'an-yen ts'ung-pao* (Salt discussion miscellany), the journal which championed the salt merchants' interests, expressed this view in commenting on the fact that for several hundred years there had never been a serious salt famine in outlying areas: "It is not the people's strength which has produced this result. It is official transportation, official supervision of distribution and the establishment of merchants by law; it is the designation of sales areas, the protection of border zones, and the limiting of times and seasons. The state has set up special officials to manage the trade, the merchants respectfully receive the law and obediently comply with it, and afterwards, we the people enjoy prosperity."[20] *Dirigisme*, then, colored Chinese views about salt reform and, at the theoretical level, this led to a negative response to Dane's program.

In practice, however, the more progressive members of all three groups found much in Dane's proposals to attract them. Among the officials, Dane received support from both metropolitan functionaries and provincial salt commissioners, who sought to escape from the hand-to-mouth finance of the early republic by reconstructing the gabelle on a more permanent basis. The reformers shared with Dane a common enemy in the salt monopolists and came to value him as a powerful ally against the monopolists' vested interests. In the past, Ching Pen-po argued, whenever there had been talk of salt reform, the salt merchants could "rely on their money to frustrate it. Now, however, there is the foreign mortgage agreement to restrain them."[21] With Dane supporting the abolition of the monopoly, the balance of power had been at last tilted against the merchants. Among the merchants themselves, while Dane was naturally opposed by the established monopolists of Ch'ang-lu and Huai-nan, he found support from forward-looking groups in Szechwan and Kwangtung, who disliked the restrictions the traditional system placed on their trade.

For each of these three groups then, common opposition to the status quo proved stronger than the theoretical differences over reconstruction, and Dane, initially an intrusive outsider in the world of Chinese salt, became a colleague whose views were adopted and whose advice was sought, even after he had left China.

Dane and the Salt Officials

The attitude of early republican officials to salt reform was colored by the progressive disintegration of the traditional institutions, and the consequent need to rebuild effective governmental machinery.

It was argued earlier that the key to Chinese politics in the decade preceding the revolution was the decline of that system of regional power-holding on which government had been based since the Taiping rebellion. This decline, for which the dynasty was partly responsible, finally engulfed the throne itself, and the revolution of 1911 completed the failure of the ancien régime. From 1912 onwards, therefore, politics were dominated by the problem of reconstruction: it was, as Ching Pen-po said, a time of state building (*li-hsien*).[22] The bases of Chinese government had to be refounded.

Adequate finance was recognized as a precondition. The Szechwan salt merchants, for example, declared in their petition that, "in establishing the republic, the first problem to be resolved is that of the government's shortage of funds." [23] Liang Ch'i-ch'ao in his preface to the first number of *YCTC* wrote: "As regards establishing a state (*li-kuo*), finance is fundamental." [24] He went on to argue that it would take time to reorganize the land tax or establish a new stamp duty, and that the customs were under foreign control; the salt tax must, therefore, for the time being at least, be the fiscal basis of the new state.

Though all agreed on the principle of utilizing the salt trade as a source of taxation, the problem of how best to do this divided not only Yüan Shih-k'ai's policy-making entourage but also salt officialdom as a whole. The dilemma confronting them may be described as the conflict between short-term and long-term policies. The traditional gabelle had broken down in the revolution of 1911, and some reconstruction was necessary if revenue was again to be secured. To create new institutions, however, would actually cost money, and, what was worse, a considerable period might elapse before they were in working order: in the meantime, the government had to survive. A tempting solution was to patch up the existing machinery to provide ready money for the government, though this policy was open to the counter objection that a true reform would thereby be delayed, perhaps indefinitely. At both levels of the salt administration — ministers and high functionaries in Peking, commissioners and subordinate officials in the provinces — there was division between those who wanted to repair the institutions of the ancien régime and those who refused to compromise the future of genuine reform. This was a difference not so much between conservatives and radicals as between opportunists and idealists, a continuation in modern form of the "impure" and "pure" streams of the old bureaucracy.

Neither group, however, had any sympathy with the principle of free trade, and both advocated a degree of interference with the trade at variance with Dane's system. The opportunists wanted to retain the traditional restrictions and quotas in order to procure a safe, if limited, revenue, while the idealists likewise opposed Dane's proposals to abolish monopoly, though for different reasons. In the past, Ching Pen-po argued, monopoly might have been abused, but this could be corrected, and the underlying principle of public control of the trade was sound. "Because of a hiccup, there is no need to abolish eating," [25] he declared in an article discussing Dane's plans. Yet in practice, in spite of their preference for state control, it was the purists who supported Dane rather than the opportunists. They shared with him a dislike of the status quo and a preference for radical solutions, while the opportunists regarded him as just another impractical doctrinaire. This alignment may be seen from Dane's relations with the two chief elements of the salt world at the official level: the heads of department in Peking and the salt commissioners in the provinces.

During the formative period of Dane's reform, from his arrival in China in June 1913 to the beginning of the revolt against Yüan Shih-k'ai in December 1915, four men held the office of minister of finance: Liang Shih-i, who was acting finance minister during the summer of 1913; Hsiung Hsi-ling, prime minister and minister of finance from September 11, 1913; Chou Tzu-chi, minister of finance from February 1914 to March 1915; and Chou Hsüeh-hsi, who held office from March 1915 onwards. The leading representative of the reformers in official circles, however, was a civil servant rather than a politician: Chang Hu, vice-minister of finance, chief commissioner of the central salt administration, and Chinese chief inspector from September 30, 1913, to June 20, 1915.

A Chekiang man, like so many of those interested in salt reform, Chang obtained the *chü-jen* degree under the old examination system and held a succession of taxation posts, among them chief of the Manchurian salt affairs bureau in 1906. He thus belonged to the small category of fiscal experts who emerged from the late Ch'ing bureaucracy. At the time of the revolution, he was associated with Chang Chien's reforming circle, and when Chang Chien found himself at odds with Chou Hsüeh-hsi and decided to return south, Chang Hu and Ching Pen-po were left in Peking as his representatives. As reported in the first number of the *YCTC*, on November 14, 1912, Yüan Shih-k'ai appointed Chang Hu Liang-huai salt commissioner, a post he held till the following July.[26]

In September 1913, Chang Hu received his three portfolios in the Hsiung Hsi-ling government. This government represented a second attempt at cooperation between the president and the Kunghotang, and the salt reformers, who were associated with this party, were strongly represented in the cabinet: the premier himself, Liang Ch'i-ch'ao (the minister of justice), and Chang Chien (minister of agriculture and commerce) were all, like Chang Hu himself, members of the Salt Society.

It was from this time that Dane's reforms began to make real progress. Dane found Chang Hu, who continued to hold his posts even after the reformers as a group left office in February 1914, a valuable and sympathetic colleague. Procedure set up to adjust differences between the two chief inspectors never had to be invoked, and in his report Dane warmly praised Chang Hu's contribution to the reform: "his far-seeing ability enabled him to appreciate the advantages of schemes and methods which were complete innovations in China, and he rendered most material assistance in the introduction of the measures, which were so successful in increasing the revenue of the Government." [27]

Dane's friendly relations with Chang Hu contrasted with the conflict which developed between Dane and the leading protagonist of the opportunist tendency, Chou Hsüeh-hsi, who was minister of finance from March 5, 1915, to April 22, 1916, and at all times in the inner circle of Yüan's advisers. Chou's opposition to Dane's reform program has already been outlined, and his reappointment as minister of finance in 1915 immediately produced a deterioration in Dane's relations with the Chinese authorities. As the chief inspector put it: "It was clear that a reactionary policy was in favour, and that the Minister was openly reverting to the methods which had been sufficiently condemned by the results of the reorganisation in the preceding 19 months." [28]

Chou's motives, however, require analysis. Although "a statesman of the previous dynasty who had obstinately retained his queue" [29] and an opponent of Ching Pen-po's type of reform, he was not a thoroughgoing traditionalist: in matters other than salt, he could be a modernizer, being, for example, the promoter of a modern cement factory at Chinwangtao and the founder of the Peking Central Hospital. However, the salt program which he had put forward in 1913, while it did not hesitate to borrow the slogans of the reformers, was designed to consolidate the position of the salt merchants. Ching Pen-po described the use of the slogans as gross misrepresentation of the real intentions of the plan, "calling a stag a horse," after the story of Chao Kao and Erh-shih Huang-ti. Chou's policy may be most accurately termed pure pragmatism. He expressed his motives succinctly in a telegram to the Szechwan salt commissioner: "There is an extremely large deficit in the budget of the Central Government for current year. It entirely depends upon Salt Revenue for relief," and he went on to argue that the reintroduction of monopoly was the surest means of restoring the income levels obtained before 1911.[30] It was this reluctance to look beyond the immediate financial needs of the government which brought him into conflict with Dane.

A similar division of personnel into opportunistic and purist groups, and a parallel antagonism and support for Dane, existed at the provincial level. The salt commissioners under the early republic varied in their attitudes. Some were frankly "selfaggrandizing officials" to use S. N. Eisenstadt's terminology, or "abusive bureaucrats" in James T. Liu's more traditional language. In Ho-tung, for example, Kao Chin-chi "arrived at Yuncheng with a retinue of dependents, more than 200 in number, for whom employment had to be found," and consequently opposed the inspectors' plans to reduce the unnecessarily large salt police and abolish the Ho-tung government transportation office. Kao remained a thorn in the inspectorate's side in Ho-tung until he was "fortunately transferred on promotion to another appointment." [31]

Most commissioners, however, as they appear from the inspectorate reports or the pages of the *YCTC,* were simply career-minded men trying to keep the system going in difficult times. Typical of these, and representative of the pragmatic stream among the provincial officials, was Yen An-lan, commissioner in Szechwan from March 22, 1914, to a date after 1916, probably August 1917.

As Duke Tsai-tse's chief adviser on salt matters and former first deputy to the department of salt administration (*yen-cheng yüan yen-cheng ch'eng*), Yen An-lan was, so to speak, part of the debris of the old regime, brought back to office by Yüan Shih-k'ai to undertake the difficult assignment of setting the Szechwan salt administration once more in motion. As the *YCTC* tactfully phrased it, he was a man who had grown old in salt affairs.[32] Yet he could not be written off as a traditionalist. In 1911, Yen had been associated with the ambitious late Ch'ing project of centralizing the salt administration under Peking's control, and after the revolution he was listed by the *YCTC* as a member of the Salt Society. Equally, however, it could not be claimed that he was a vigorous reformer. According to Ching Pen-po, he had been an opponent of Chang Chien's before the revolution, and his later pronouncements as recorded in the *YCTC* do not show him to have been a radical by the standards of 1913.

After he had been in office a few months, Yen called a meeting (reported in the seventeenth number of the *YCTC*)[33] of 200 leading salt merchants to explain his policies. His approach was traditionalistic. He opened with a résumé of salt administration history from the Wan-li period onwards, emphasizing the difference between the *kang* and *p'iao* systems: the *kang* system was feudal (*feng-chien*), the *p'iao* system imperial (*chün-hsien*). He went on to criticize the taxation-at-source system (*chiu-ch'ang cheng-shui*), then being practiced in Szechwan. It was not, he said, in fact like Liu Yen's system, and anyway Liu Yen was not a model to be imitated: he was like the Han dynasty general Li Kuang, presumably meaning by this that he was an individual genius who did not found a tradition. From there Yen went on to outline his own program. It amounted to a restoration of the traditional *kuan-tu shang-pan* system: privileged merchants, designated distribution areas, and sales quotas. Nevertheless, the plan did show evidence of current thinking in three respects: the privileged merchants were to be companies, not individuals; the units of the monopoly were to be large segments of the division, *an* (lit., "ports"), not the traditional hsien; and a serious attempt was to be made to tax land-carried salt through the so-called *kung-yüan* or licensed marketing agencies. It was essentially an amalgam of traditional and modern ideas designed to meet the government's immediate need for funds.

Given such an outlook and such a program, Yen's relations with Dane were bound to be full of friction, even though the chief inspector conceded that Yen conducted the administration "honestly and well upon the old lines." Difficulties arising from the coordinate jurisdictions of the commissioner and the inspectors were the first reason given by the *YCTC* for Yen's frequent threats of resignation. At the same time, Dane had to fight a long and ultimately successful battle to abolish Yen's privileged companies' monopoly over waterborne salt in Szechwan. The commissioner's scheme, he admitted, was "well calculated to secure for the Government with comparatively little difficulty in times of peace a considerable revenue"; but it sacrificed future growth, "the legitimate expansion of revenue, which might reasonably be looked for in times of prosperity." Moreover the principle of monopoly did nothing for the consumer: "Under the scheme also the interests of the millions of people in Szechuan and Hupeh, who consume Szechuan salt, were entirely sacrificed." In these circumstances, Dane and Yen could not but be at cross-purposes.[34]

At the other end of the scale from Yen were a group of officials, such as Hsiao K'un in Yunnan or Liu Hung-shou in Fukien, who advocated more ambitious programs of reconstruction and were less concerned with immediate results. Dane did not always agree with the principles on which these programs were based, but he was able to establish a measure of collaboration with their authors.

The Yunnan commissioner's program, which was published in the nineteenth number of the *YCTC* and is also referred to in Dane's report, involved the closure of outlying wells, resettlement of workers in the main centers, and the construction of government depots for storing salt. This intensification of control was in line with Dane's plan, and two other proposals of the commissioner, the abolition of separate consumption areas for the three groups of wells and the introduction of uniform rates of tax for all wells, were adopted. Other items, such as reforestation and government moves to improve evaporating methods, though rejected by Dane as overambitious, were later revived by the inspectors. The only difference of principle was over free trade: the commissioner, taking the interventionist standpoint characteristic of Chinese salt experts, wanted government transportation to the border areas and preferential rates of duty to maintain Yunnan sales, while Dane favored the free import of Kwangtung, Tongking, and Burmese salt.[35]

In Fukien, Liu Hung-shou advocated a complete government monopoly of purchase, transportation, and distribution. Dane described him as "a man of considerable capacity and energy." His system was anathema to the chief inspector, but, because of its popularity with the Chinese reformers, "the monopoly was allowed a fair trial," as regards the internal trade of the province. Salt was purchased by government agency at the works south of Foochow, brought to depots at the capital, and then transported up the Min River to 464 government sales offices. The difficulty was that Fukien was a salt-surplus province, and the merchants were tempted to channel the export salt back into the accessible home market. Infringement of the monopoly was hard to prevent, and the monopoly's only remedy was limitation of production, which the inspectors feared would harm bona fide exports to Swatow.[36] In 1917, therefore, Dane ordered the gradual dismantling of the monopoly system, though he retained for a time the system of government transportation as far as Foochow. The case of Fukien shows that Dane was prepared in certain circumstances to give the Chinese reformers their head, and to adapt his ideas to theirs. His sympathies lay with them rather than with their less idealistic but perhaps more practical colleagues.

Dane and the Salt Merchants

In China the term salt merchant was used of at least seven groups operating the salt trade. Closest to the salines were the manufacturers (*ch'ang-hu,* lit., "yard households") who might be millionaire capitalists as in Szechwan, "people of means" as in Ch'ang-lu, or mere artisans subject to more powerful firms as in much of Huai-nan. Next, were the so-called yard merchants (*ch'ang-shang*), men of great resources always who might, as in Huai-nan, organize production, but who more generally, like the *ao-shang* (lit., "granary merchants") in Liang-che, acted as export brokers between producers and distributors.

Third, there were the monopoly merchants (*yin-shang* or *p'iao-shang*) who held the rights of purchase and supply from specified yards (*ch'ang*) to designated hsien or parts of hsien. Some, as in

Huai-nan, were active traders; others, as in Ch'ang-lu, "sublet their rights of transportation and lived, like absentee landlords, upon the rents which they received." [37] In this case, there then appeared a separate group of transportation merchants (*yün-shang*), though this term could also be applied to the monopoly merchants (*yin-shang*), who were themselves active businessmen.

Next, at the provincial distributing centers such as Ta-t'ung, Shasi, and Hankow were the import brokers (*hang-shang,* lit., "guild merchants") or salt hongs, who purchased the salt for resale, or arranged its resale, to the provincial distributors, who themselves went by a variety of names: in Hupei, *p'u-fan* (lit., "store merchants") for the urban areas and *shui-fan* (lit., "water merchants") for the backcountry; in Huai-pei, *hu-shang* (lit., "lake merchants"); in Szechwan *an-shang* (lit., "port merchants"). These in turn sold to the public, either through their own shops or through petty retailers and peddlers.

Finally, interlacing these activities were the various ancillaries concerned primarily with transportation: the canal junkmen who in Huai-nan took salt from the works to Shih-erh-wei; the river junkmen with larger ships who "controlled the carrying trade on the Yangtze from Shiherhwei upward, as far as to Yochow in Hunan"; or in Szechwan, the "small but avaricious coterie of boat-brokers" who controlled the 5,000 specially built vessels which carried Tzu-liu-ching salt the sixty miles to the mainstream of the T'o River.[38]

All these groups were active in defending their privileges. The manufacturers of Lo-shan in Szechwan, for example (who in this case acted also as transportation merchants), resisted the inspectors' proposal to open their exclusive markets in northwest Szechwan to salt from the neighboring center of Chien-wei. At Shasi in 1922, when "the Foreign Assistant District Inspector requested the Chief Inspectors to abolish the Shasi Salt Brokers' Combine," the merchants concerned invoked the support of the local warlord, marshal Hsiao Yao-nan; in 1913, on the other hand, the civil governor of Hupei, "in response to petitions from over 180 salt shops," had abolished the monopoly of the Hankow salt brokers. In Hunan in 1920, a proposal by the chief inspectors to transport salt by steamer instead of by junk "could not be carried out, owing to the opposition of the Yangtze boatmen corporations." [39] Of all the groups, however, it was the *yin-shang,* whose monopoly was best defined and whose rights were most deeply entrenched, who were most active in resisting outside encroachment in the period of the foreign inspectorate.

Until the revolution of 1911, there was no national organization claiming to represent the whole of the salt trade, nor even one level of it. Business, like much else in imperial China, was regionalized: there were only provincial or local guilds, such as the Ho-tung salt makers guild, which fixed prices at the An-i-hsien salt lake, or the Huai salt company of monopoly merchants and salt hongs, who transacted business with the officials at Hankow.

After the revolution, the development of a national salt reform movement directed against the salt merchants forced them to organize nationally in return, and in the winter of 1912–13, a federated salt trade association (*ko-sheng yen-yeh hsieh-hui*) was set up at Shanghai. The main activity of this body was the publication of a magazine, the *T'an-yen ts'ung-pao* (Salt discussion miscellany; hereafter *TYTP*), designed to counteract the propaganda of Ching Pen-po's *YCTC.* The new journal was a monthly, and it ran only for eighteen consecutive issues, from April 1913 to September 1914, publication then being suspended until 1930.[40] At the outset, the leadership of the movement was in the hands of the monopoly merchants of Liang-che, Ching Pen-po's original target. In 1914, however, leadership shifted to Ch'ang-lu as a result of Dane's attempt to introduce free trade in that area. The Ch'ang-lu *yin-shang* organized a conference of salt merchants in Peking, which, according to a report in the *YCTC,* met in offices in Shun-chih-men Street and was attended by representatives from Ch'ang-lu, Liang-huai, Liang-che, and Kwangtung. Dane's program and that of the Chinese reformers were both condemned, and a petition was presented to the prime minister Hsü Shih-ch'ang calling for the maintenance of the status quo.[41]

The Salt Trade Association was primarily the creation of the *yin-shang,* those most threatened by reform, though it was probably supported by other levels of the salt trade which stood in a client relationship to the *yin-shang.* The outlook of the association can be gauged from the articles of the *TYTP.* These indicate that, while the merchants shared the *dirigiste* bias common to all contemporary Chinese salt experts, they were more effective in criticizing than in putting forth constructive proposals. Their attitude was particularistic: they raised practical objections to the plans of others but could articulate no general formula of their own. In particular, the *TYTP* advanced cogent arguments against the taxation at source plus free trade system (*chiu-ch'ang cheng-shui tzu-yu fan-mai*) and against Ching Pen-po's formula of *chiu-ch'ang chuan-mai* (lit., "monopoly at the yard").

Taxation at source, it was argued, was applicable only in special circumstances, such as Ho-tung, Szechwan, or Hua-ting, but not to the country as a whole, since the administrative costs of control would be too great.[42] The existing salt works magistracies (*ch'ang-kuan*) could not perform the task unaided, since the areas they covered were too extensive for the number of personnel employed; nor could the job be done by the higher level salt offices at commercial centers such as Hsi-pa on the Grand Canal, since these were too far from the salt fields. New institutions would have to be set up, complete with officials and clerks, increasing administrative costs many times over, and even then the problem of preventing smuggling would still be formidable. So much for taxation at source.

Pertinent anti-laisser-faire arguments were deployed by the merchants against the free-trade half of the program: it was only the monopoly, they said, which kept the salt business profitable in outlying areas and throughout the trade prevented excessive competition and loss of profit margins. Without the monopoly, "capital will yield a loss and in the course of time, transporters will regard the salt business as a dangerous and difficult road." In short, overcapitalization would be succeeded by investment deficiency, and the end-product would be an uncontrolled monopoly of local vested interests, who would exploit the situation to their own advantage, "buying cheap and selling dear." [43]

Against price equalization too, the salt merchants' arguments were generally practical.[44] It would be difficult to fix a uniform purchase price for salt: what would be fair to the manufacturer would not be fair to the consumer. The attempt to equalize retail prices would also cause trouble, since it would involve upsetting an established and accepted price structure: what would be fair to the 100-cash-a-catty areas would be unfair to the 30-cash-a-catty areas.

The reformers' idea of replacing artificially evaporated salt by solar-evaporated salt (to standardize the product) was worst of all, for only Fengtien, Ch'ang-lu, Shantung, and Huai-pei produced *shai* salt. The reformers' plan thus entailed rooting up the salt industry in the rest of China, which was, the merchants protested, both inhuman and imprudent. There would be disorder and rebellion on the precedent of Chang Shih-ch'eng, the salt bandit and rival of Chu Yüan-chang at the beginning of the Ming dynasty.

If many of these criticisms seem well aimed, the salt merchants found it harder to frame a positive program, acceptable to all of them. The article which criticized *chiu-ch'ang chuan-mai* so convincingly concluded lamely by stressing the diversity of circumstances and institutions in China, and advised the ministry of finance to call a conference of salt experts. The meeting called in Peking by the Ch'ang-lu merchants immediately split into three factions. The merchants of Ch'ang-lu and Huai-pei, where abolition of the monopoly was either an accomplished fact or imminent, were less hostile to change than those in Huai-nan and Liang-che, where the ancien régime still stood intact, while the merchants of Kwangtung, where monopoly had been abolished as early as 1911, were more interested in the revival of tax farming than in commercial privileges. It was these differences, perhaps, which caused the *TYTP* to cease publication in September 1914 after only eighteen issues. The merchants were not basically interested in overall reform programs. The *TYTP* and its parent body had proved unable to generate a united platform or a common front among the members. In this they contrasted with the *YCTC* and the Salt Society of the reformers.

The merchants were prepared to be active, however, over the particular problems of their own region or branch of the business. These local interests were well documented by the *TYTP:* one finds reports of a protest by the Tzu-liu-ching manufacturers against fraud by the boatmen, or of a petition of the Nanchang salt guild to the Kiangsi military governor against a price increase, for example.

Another example of this genre, and an indication of merchant attitudes towards reform, was a petition of the All Szechwan Well Merchants Association (Szu-ch'uan ch'üan-sheng ching-shang kung-so) to the provincial assembly in 1913.[45] The occasion was the meeting of parliament in Peking and the proposed enactment of a national salt law, and the petition was a formal submission by the Szechwan manufacturers on the policy they wanted their representatives to pursue.

The petition opened with an attempt to link local salt problems to wider issues: finance was the paramount problem in the republic and it could be solved only if trade flourished; in Szechwan, because of the undeveloped character of the economy, salt was the chief staple of business; good government therefore depended on an expanding salt trade. Two measures were necessary to ensure this. First, tax differentials must be maintained, both between different producing centers within Szechwan, and between Szechwan and other kinds of salt. Szechwan production costs were higher than those of other regions because of the expense of sinking wells, and fuel costs were higher in some Szechwan areas than in others. If an equilibrium was to be maintained both within and without, then the retention of the differentials was vital.

Second and even more vital was the preservation of Szechwan's export markets in Hupei, Kweichow, and Yunnan, without which the province would face a serious crisis of overproduction, and wells would have to be closed. The petitioners, therefore, asked the assembly to secure an official confirmation of their traditional markets. The petition closed, as it had begun, on a political note: these two measures between them would safeguard the two traditional desiderata of salt policy, the finances of the state and the people's livelihood (*kuo-chi min-sheng*).

This petition, skillfully defending local vested interests in terms of traditional political morality, was representative of merchant attitudes on the eve of Dane's reforms. The Szechwan well merchants were unusually articulate, but other sources indicate the same preoccupation with particularistic issues. The salt interest at the merchants' level was a fragmented one, and it reacted to Dane's reforms in a fragmented manner, depending on how its sectional interests were affected.

Thus, in Ch'ang-lu the monopoly merchants vigorously resisted Dane's attempts to introduce free trade: "The prospect of the cancellation of the merchants' Yin rights caused therefore considerable excitement. Mr. Chang Hu had a heated discussion with the merchants in Tientsin; and a violent anti-foreign manifesto appeared in a Tientsin local paper." The president was petitioned, and the merchants "raised a large fund to defend their privileges." An article in the *TYTP* was sharply critical of Dane for illegitimately extending his authority and interfering in China's domestic arrangements: "What still worries us in this matter is that it was a telegram from Dane which effected the abolition of the Chihli–Honan *yin* laws." [46]

In Kwangtung, on the other hand, the merchants needed Dane's support against the authorities. The central government, probably because of the south's tendency to independence, disliked Kwangtung salt and tried to exclude it from Hunan through an additional levy on the provincial border. Dane, regarding Kwangtung as the "natural source of supply" for western Hunan and eager to relax the barriers preventing competition, favored the Kwangtung claims, and he "strongly urged that this high taxation of Kwangtung salt in the adjacent Provinces was inexpedient and inequitable." His antimonopoly policy, too, was popular in Kwangtung: in 1921, when the Kuomintang salt commissioner Tsou Lu attempted to restore the monopoly system, "the proposed scheme was frustrated in consequence of the strong opposition of the Provincial Assembly and the salt merchants." [47]

Although the reaction to Dane was conditioned by particular, sectional interests, it was not haphazard but followed a definite pattern: support came from those merchants who, in some sense, stood for economic development, and opposition came from those content with the traditional equilibrium. This may be seen from the case of Szechwan.

The salt interest in Szechwan was unusually complex. Although the merchants might present a united front to the provincial assembly in their petition, they were deeply divided among themselves. The most acute of these divisions was that between the Tzu-liu-ching area on the one hand (often known as Tzu-kung, from Tzu-liu-ching and its neighbor to the west, Kung-ching), and the other producing centers, in particular Chien-wei on the Min River. The basis of the division was fear that, given open competition, Tzu-liu-ching salt would drive out other kinds: "Tzekung enjoys many natural advantages which enable these works to supply the cheapest salt. Many wells at Tzekung supply gas . . . All other works use either coal or wood as fuel . . . Further, Tzekung wells furnish better brine, are much deeper and can supply in a given period much more brine than the other works." [48] To some extent, this division corresponded to that between the eastern and western halves of the province, centered respectively on Chungking, the leading Yangtze port, and Chengtu, the terminus of the imperial highway from Sian.

One of the functions of the traditional monopoly had been to neutralize Tzu-liu-ching's advantages and redress the balance in favor of the smaller centers, but despite this, in the thirty years between Ting Pao-chen's reforms and those of Dane, Tzu-liu-ching salt had been gaining on its competitors and this lead was accelerated by the introduction of free trade in 1911. In 1882, the Szechwan salt laws gave Tzu-liu-ching a total quota of 827,854 piculs, while its principal rival Chien-wei received a quota of 1,270,088 piculs.[49] In 1904, Sir Alexander Hosie was told that the annual output of salt at Tzu-liu-ching was 2,744,000 piculs, while that of Chien-wei was only 838,723 piculs. In 1914, the *TYTP* reported a petition from the Chien-wei town council to the provincial assembly, which complained that Tzu-liu-ching salt was encroaching on Chien-wei's traditional preserves in northeast Yunnan. In 1915, when Yen An-lan revived the monopoly, he allocated Tzu-liu-ching a quota of 2,332,160 piculs for water transportation and gave a quota of only 849,920 piculs to Chien-wei; these figures were later revised to 2,722,840 and 717,600 piculs.[50]

As a result, fear of Tzu-liu-ching increased rather than diminished. When Yen An-lan went to Szechwan in 1914, he put forward a scheme for organizing the salt industry through a parent company at Tzu-kung and subsidiaries at the other main centers.[51] Whatever Yen's intentions, and given his conservative outlook he probably intended only an adjustment of the traditional quotas, the formation of the Fu-jung salt trading company (Tzu-liu-ching was in Fu-shun hsien; Jung-hsien was a neighboring salt district) produced a storm of controversy, which can be followed in the *YCTC*. The All Szechwan Merchants' representatives telegraphed their fellow provincials who were officials at the capital, asking for their support against the Tzu-liu-ching deputy Li Hsin-chan, who was scheming "to control at will the whole Szechwan salt business." A supporter of the Fu-jung company replied, accusing the other side of "falsely pretending the public interest in order to help their own private interest." He rebutted the suggestion that Tzu-liu-ching sought to monopolize the trade, claiming that it sought only to preserve its own markets. This was followed by yet another telegram from the anti-Tzu-liu-ching party.[52]

By now, Dane had entered the scene, and the ground of the controversy had shifted. In the final version of Yen An-lan's scheme,[53] the existing system of free trade in waterborne salt was to be replaced by eighteen monopoly companies, of which six were allocated to Tzu-liu-ching and four to Chien-wei, and controversy turned on the size of the quotas to be allotted to each center. The Tzu-liu-ching merchants felt their quota was too small: as Dane put it, "The Commissioner has altogether underestimated the production at Tzeliuching, and the scheme of allotment is unduly favourable to Chien Chang." One of the advantages claimed by Yen An-lan for his system was that it would facilitate increased taxation, but the Tzu-liu-ching merchants petitioned, offering to pay a higher rate of tax voluntarily if the status quo was maintained: "These men," said Dane, "who claim to represent 9/10 of the Tzeliuching brine and 7/10 of the gas interests, invite Government to impose a uniform duty of $2 a picul if free trade is allowed to continue." In championing free trade, then, Dane was also championing the low-cost, technically sophisticated Tzu-liu-ching interests against Chien-wei and the more conservative remainder of the province.[54]

The abolition of the monopoly in 1916 benefited Tzu-liu-ching at the expense of Chien-wei: in 1919, its turnover was 3,415,885 piculs, and there would have been a further 600,000 if conditions in Hupei had been normal; Chien-wei, on the other hand, handled only a total of 452,815 piculs in 1919, despite this being a reasonably good year for its export trade to Yunnan and Kweichow. Tzu-liu-ching contributed more than half the total of 6,105,763 piculs of taxed salt in Szechwan in 1919. It also continued to be more technically advanced than the other centers: in 1922, fifty steam engines, "used for raising brine from the wells," were reported in operation, arrangements had been made to ship salt down the Yangtze gorges by steamer instead of junk, and a railway from Tzu-liu-ching to Lu-chou was under consideration. No such developments were reported from any of the other manufacturing areas.[55]

Dane and the Salt Reformers

The Salt Society had been founded at the end of 1912 by Chang Chien and Ching Pen-po, with the aim of influencing public opinion in favor of thoroughgoing reform and against the conservative policy of Chou Hsüeh-hsi. The first meeting was held in Peking on December 7, 1912, and the first number of the *YCTC,* containing Chang Chien's reform plan, appeared shortly afterwards. The magazine continued to act as a platform for the reformers and to provide a stream of information and comment on different aspects of the administration until 1915, when Ching Pen-po was forced to cease publication because of his refusal to use Yüan Shih-k'ai's reign title in the dating of the journal.

The reformers' program involved nontraditional centralization, standardization, and rationalization: like Dane's own plan, it aimed at a national administration, it sought to control all salt and tax it at source, and it wanted to replace tax farming with direct administration.[56]

Unlike Dane, however, the reformers were not attracted by the principles of laisser faire, and the principles on which they proposed to build a new system were pronouncedly *dirigiste.* Ching Pen-po was explicit on this point. In an article in the *YCTC* comparing his own system with that of taxation at source, he stated that while taxation at source supported laisser faire (*fang-jen chu-i*), his system favored interventionism (*kan-she chu-i*).[57] Basically, the reformers lacked Dane's confidence in the market. Their contrasting attitudes towards the question of inducements to invest illustrate this. Dane stated his view in a letter to Hillier, the consortium representative, on April 30, 1914: "Advances to merchants for salt transportation will be money thrown away. There are numbers of merchants only too eager to transport salt without advances, if they are allowed to do so"; and again in his report, "Experience has shown also that instead of salt merchants requiring assistance to enable them to transport salt, any number of merchants are only too ready to transport it." [58]

The Chinese reformers, on the other hand, believed that additional incentives were required to achieve an adequate level of investment. Ching Pen-po argued: "If zones for the circulation of salt are not mapped out, then the merchants will concentrate on local markets and avoid long distance enterprises. In places near the *ch'ang,* then, there will be glut, while in remote and out of the way places, there will be high prices and salt famine." He gave the situation in Szechwan consumption area in the previous year as an example: since the introduction of free trade, there had been a glut of salt at Ichang and Shasi, and an incipient salt famine in Kweichow. Yet under the old monopoly system, trade to Ichang and Shasi despite its good water communications had not expanded, while the Kweichow trade despite its poor land routes had not contracted. The monopoly and the quotas were required to readjust the natural imbalance of the economy, which otherwise would not work to an optimal result: "A merchant really seeks profit. If he knows for certain there is profit, then he will begin to venture and take risks." [59]

Other examples can be found of Ching Pen-po's distrust of the uncontrolled working of the economy: fears that "supply and demand will not correspond," [60] fears of overproduction leading to massive smuggling, fears that free trade will only be a cloak for a new, unsupervised monopoly of big merchants, fears that foreign merchants will rob the Chinese of their trade. All this, whether one regards it as the failure of traditional thinking to grasp fully the significance of the market as a control mechanism, or as intelligent anticipation of Keynesian economics, is in marked contrast with Dane's Victorian confidence in the workings of economic nature.

The strong presumption in favor of government intervention in the salt administration is evident at every stage in the reformers' thinking, and the leading document of the movement, Chang Chien's *A Plan for the Reform of the National Salt Administration,* bears this out. The following discussion will analyze the long- and short-term implications of the reform plan and its administrative antecedents.

Although the plan was published under Chang Chien's name, the *YCTC,* in its accounts of the genesis of the movement, makes it plain that the reform plan was in fact a joint compilation, and that Chang Chien, Chang Hu, Liang Ch'i-ch'ao, Hsiung Hsi-ling, and Ching Pen-po collaborated in its composition, with the last making the greatest contribution.[61] Chang Chien's own preference had been, until 1912, the policy known as "taxation at source and minimum interference thereafter" (*chiu-ch'ang cheng-shui jen ch'i so chih*). Late in 1912, however, possibly as a result of his recent experiences as Liang-huai salt commissioner, and possibly to secure unity among the reformers, he agreed to adopt as the ultimate objective of the reform platform the formula advocated by Ching Pen-po, *chiu-ch'ang chuan-mai* (lit., "monopoly at the yard").[62]

Ching's own schema explains this formula. In the first number of the *YCTC,* Ching Pen-po suggested that all official salt administrations could be divided into taxation systems (*fu-shui hsi*), in which the government confined its activities to levying tax, and business systems (*ying-yeh hsi*), in which government control extended to commercial operations as well. He placed the taxation-at-source system under the first heading, and the various monopoly systems (*chuan-mai chih*) under the second, the implication being that monopoly involved more state control than taxation at source. Monopolies, however, could be further qualified as to whether they were complete (*kuang-i chuan-mai*) or partial (*hsia-i chuan-mai*), direct (*chih-chieh chuan-mai*) or indirect (*chien-chieh chuan-mai*).

This categorization determined the content of the reformers' phrase: there is reason to believe that by *chiu-ch'ang chuan-mai* the reformers signified an extended monopoly, one both complete and directly operated by officials. The principal example of the system as a going concern was the Kirin–Heilungkiang office set up in 1908. This was described in the Chinese version of Dane's report on Manchuria as "official acquisition, official transport, official sales" (*kuan-shou, kuan-yün, kuan-mai*). Several of the reformers had been connected with this development, for example Hsiung Hsi-ling and Chang Hu, and their plan explicitly declared that "In Manchuria, the Government monopolistic control is direct and complete, and this is the best part of the present Salt Administration." Dane certainly believed that "A complete Government monopoly of the purchase, transportation and sale of salt was also the ideal of Chinese salt experts." A similar view was held by the reformers' opponents, the merchants: a writer in the *TYTP* argued that the reformers' plan would lead to "official acquisition, official transport" (*kuan-shou kuan-yün*).[63]

Chiu-ch'ang chuan-mai, then, meaning complete monopoly from production to retail, was the ultimate objective of the reformers' program. Their immediate aims, however, as put forward in the plan, did not go as far as this. Complete monopoly would involve both buying out the capital equipment of the salt manufacturers (requiring, according to the *TYTP,* between 10 and 20 million taels) and supplying the capital for transportation and retail, and the reformers shrank from such an outlay. Instead, they advocated a monopoly which was both partial, in the sense of not encompassing the whole of the trade, and indirect, in the sense that some parts of the monopoly were not to be directly operated by officials. This interim proposal was summarized in the formula "popular manufacture, official acquisition, merchant transport, popular sales" (*min-chih, kuan-shou, shang-yün, min-mai*).[64]

This sounds like the traditional *kuan-tu shang-pan* system in a new dress. The reformers insisted, however, that their system would introduce a fundamental change by restoring to the state powers which had been allowed to lapse to the merchants. The existing system left the merchants supreme in the monopoly, but, argued Ching Pen-po, "Mr. Chang's plan in all respects takes the state as the basic premise, regards the state as the monopoly's theme. The so-called merchant transportation it regards as only part of the business." [65]

The key element in the reformers' plan was "official acquisition" (*kuan-shou*). This would enable the state to make socially desirable adjustments at any time: "regulation of production, equalization of salt prices, establishment of production quotas, all pivot on official acquisition." At the same time, it avoided the overextension involved in direct monopoly: "To take the major salt *ch'ang* and have them repurchased by the state and placed under direct government management is unthinkable, but to acquire the salt and not acquire the production, i.e. people manufacture, official acquisition, is not unthinkable." [66]

Kuan-shou was not a new term. Tseng Yang-feng in his standard history of the salt administration uses it to describe monopoly purchase in the salt system of the *Kuan-tzu* and Liu Yen's program under the T'ang, but the reformers gave it a new sense. All salt must be sold to government agencies at the works, which would then fix the price, taking into account both production costs and an appropriate return on producers' capital. Next these agencies would sell the salt to the transportation merchants at a nationally uniform price, to be arrived at by averaging the purchase prices and adding a uniform salt tax. The transportation merchants' price to the retailers was also to be fixed by law, so that the price ultimately paid by the consumer would be differentiated only by transportation costs, and not, as heretofore, by different production costs and different rates of tax as well. "Official acquisition" was thus to act as a form of price equalization.[67]

Kuan-shou was also to be instrumental in reorganizing the pattern of manufacture. Under the old regime, the merchants were involved in a dual monopoly: they had to buy their salt at designated manufacturing centers, and they had to sell it in prescribed areas. Under the reformers' scheme, the merchants were still to possess their tied markets (though these were to be arranged in larger units than before and were to be owned by companies, not individuals), but they were to be free to buy their salt from any of the government agencies which had acquired it from the manufacturers. Salt from different areas and perhaps produced by different methods would thus compete for a single market, though the competition would be one of quality, not price.

In this way, official acquisition would further an objective not explicit in the reform plan, but evident from the *YCTC* and from criticism in the *TYTP:* the eventual replacement of artificially evaporated salt (*chien-yen*) by solar-evaporated salt (*shai-yen*).[68] The reformers disliked production by more than one method: "The difficulty of abolishing the *yin* divisions is due to the impossibility of making the salt tax uniform, which in turn is due to the impossibility of making the sources of salt uniform . . . it is a major hindrance to the salt administration."[69] Abolition of *chien-yen,* however, would not only facilitate tax and price equalization, it would also lower prices since *shai* salt cost only 2–3 cash a catty to produce, while *chien* salt might cost as much as 7. Further, this development would improve the quality of salt available: the *chien* method could, it is true, produce the better salt, but its manufacturers were forced to adulterate it to keep its price competitive, so in practice it was the inferior product, kept in business only by its tied markets. By means of increased state control and the operation of *kuan-shou,* the reformers thus offered the consumer cheaper, better salt at a uniform price.

The salt reform plan as a whole was a collective composition, but the idea of an agency to equalize prices by buying all salt on production was Ching Pen-po's. The term *kuan-shou* used in this sense appeared first in the program of the Chekiang salt bureau (*yen-cheng chü*) of which Ching was secretary: "if we want to unify the salt price, we must begin with *kuan-shou.* All salt, from whatever *ch'ang* and produced by whatever method, shall be acquired at a uniform price fixed by official agency." [70]

Ching does not enlarge on the intellectual influences which prompted this innovation, but conjecture is not difficult. In the first place, Ching would have been familiar with the concept of the ever-normal granary (*ch'ang-p'ing-ts'ang*): indeed, he suggested the establishment of such institutions for salt as a possible alternative to direct government transportation in remote places where merchant investment would be inadequate. It must be admitted, however, that intermittent price stabilization on a local basis is at most only cognate to permanent, nationwide price equalization, even though both involved the principle of government marketing to control prices.

Second, Ching was certainly influenced by contemporary Japanese practice: in an argument designed to show that the taxation-at-source system would be unworkable in a country where the rate of tax was as high as it was in China, he pointed out that the Japanese gabelle had switched from a taxation system to a monopoly system when their rate of tax was raised after the Russo-Japanese war. In the reform plan itself, Japan was cited as the example of a limited monopoly, the kind Ching was advocating: acquisition (*shou-na*) and transportation were in the hands of officials, while manufacture and retail were left to merchants. Ching also commended Japan for establishing a special agency for salt affairs and not working through the finance department, and selected Japan as the first foreign country for investigation in the *YCTC.*[71]

Nevertheless, there is reason to believe that Japan was unlikely to have been the main influence on Ching's thought. The new Japanese salt law dated only from 1905, while Liang Ch'i-ch'ao in his preface to the first number of the *YCTC* (1912) stated that Ching Pen-po had been studying the salt administration for more than ten years; and Ching himself, in his history of the reform, says that in the summer of 1903 he was already advocating *chiu-ch'ang chuan-mai* against the taxation-at-source system supported at that time by Chang Chien. Nor were all the references to Japan in Ching's writings complimentary: in an article against foreign interference in the salt administration, he cited the example of Taiwan where the Japanese acquisition agency gave the producer a lower price than he would have received on the mainland of China.[72]

The main influence on Ching, it may be suggested, was that stream of administrative thinking about salt problems which originated with Ting Pao-chen (see above, Chapter 2), and which continued to influence the work of imperial officials such as Chao Erh-hsün and Hsü Shih-ch'ang. Although Ting died in 1886, his principal collaborator in reforming the Szechwan administration, T'ang Chiung, was in office as head of the copper administration as late as 1906, publishing his collected writings in 1908. This school stressed the official component of the salt monopoly against the quasi-liberalism of T'ao Chu, and still more against the taxation-at-source thesis of Ku Yen-wu. It belonged to the legalist end of the Confucian spectrum and emphasized the need for resolute official action, particularly in border regions such as Szechwan and Manchuria where many of its exponents held office.[73] These attitudes were inherited by the reformers and carried over by them into a modernization program that embodied both an attack on the traditional system and an extension of state power.

Given this outlook, the Chinese reformers' attitude towards Dane would depend on whether the course of events emphasized their common opposition to the tradition or, per contra, the difference between the systems they proposed to put in its place. In fact, the relationship, which may be traced in successive numbers of the *YCTC,* developed through three stages: initial hostility, growing appreciation, and incipient cooperation.

The hostility of the reformers was aroused as soon as foreign control of the gabelle was mooted during the course of the reorganization loan negotiations in 1912. Indeed, according to the *YCTC*'s account of the first meeting of the Salt Society, opposition to foreign control was one reason for the formation of the society, along with the desire for internal reform though Ching Pen-po in his later history of the reform movement chose not to emphasize this aspect. Besides nationalist antipathy to any loss of sovereignty, the reformers had a number of specific objections to foreign control.

First, they believed that because of the ramifications of the salt administration, the loss of sovereignty could not remain strictly limited as in the case of the Maritime Customs and the post office: suppressing smugglers involved control of military and naval forces; fixing quotas involved the power of the census; improving transportation routes involved interference with rivers. The result would be the "Egyptianisation" of China, colonialism under the guise of debt guarantees.[74]

Second, they argued that reform under a foreign aegis was bound to be ineffective. Even Chinese could not understand the complexities of the salt administration, how much less foreigners. Antiforeignism would reinforce the salt smuggler, and the result would be widespread unrest.

Third, at best the foreigner would maintain the status quo. Indeed, the reformers believed that Chou Hsüeh-hsi had included "the superintendence of the issue of licenses" [75] among the powers to be granted to the foreigner under the reorganization loan agreement specifically to give the salt merchants' *yin* and *p'iao* the status of entrenched treaty rights and hence make Chang Chien's plan, which involved their abrogation, impossible for the duration of the loan. After the signing of the agreement, the *YCTC* reported, Chou Hsüeh-hsi was said to have declared: "What does it matter how angry Chang Chien is? Within the next 39 years, no matter what changes may occur in the political situation and the constitution, there is no loophole which can give rise to the abolition of the *yin*." [76]

Finally, if the foreigner did succeed in introducing reforms, the most likely was the free import of foreign salt to facilitate control of the salt administration by the Maritime Customs or an analogous institution. Chinese salt was not competitive and would eventually be driven off the market, destroying the livelihood of the million-odd salt households and all the trading interests. This was not as fantastic a suspicion as it might appear: in Bengal, as a result of British rule, the native salt industry had been destroyed by imports from Europe, Egypt, and the Red Sea, and in Ch'ing times the imperial government had taken care that the import of salt into China be prohibited in the treaties.[77] The reformers were not altogether unreasonable to fear that economic "Bengalisation" might parallel political "Egyptianisation."

Dane's arrival in China did little to dissipate these misgivings. His public adoption of the policy of taxation at source plus free trade, which was already under attack by the Chinese reformers, exacerbated their hostility. However, signs of a thaw soon appeared. Dane had several interviews with Ching Pen-po in which he promised not to reach a final decision on policy before further investigation, and, in an interview with a Japanese newspaper, he repeated that he had no cut and dried reform plan and expressed proper respect before the complexity of the administration.[78] This conciliatory attitude was reciprocated: the *YCTC* began publishing Dane's memoranda on the various salt divisions verbatim, and expressed cautious optimism for the future.

Détente was fostered by the growing opposition of the salt merchants to reformers, both Chinese and foreign. In the early part of 1914, Dane was endeavoring, ultimately unsuccessfully, to abolish merchant monopoly in the Ch'ang-lu salt division. The salt merchants' conference called by the Ch'ang-lu monopolists attacked both Dane's *chiu-ch'ang cheng-shui* and Ching's *chiu-ch'ang chuan-mai,* and the agitation became sufficiently serious for Yüan Shih-k'ai to intervene personally by convening a conference to settle the matter "without provoking the same general unrest as resulted from the proposal to nationalize the country's railways." [79] Hostility to the *yin* system had always been one of the basic elements of the Chinese reform movement. Indeed, in one respect, the reformers were more hostile than Dane, for while he was prepared to accept the merchants' right to compensation for the extinction of their monopolies, Ching Pen-po refused on the ground that the *yin* were not a genuine property right. The prominence of the *yin* question in the summer of 1914 made the reformers more sympathetic towards Dane, whom the *YCTC* recognized as "the driving force behind the proposal to abolish the *yin*." [80]

The new attitude was expressed in an editorial in the *YCTC* which summarized and commented on Dane's own articles in the journal on the history of Indian salt administration.[81] The publication of these articles was in itself a sign of greater appreciation, and, as Dane's policy was explicitly based on his Indian experience, their appearance gave the magazine an opportunity to reassess its relation to that policy.

The magazine discussed only two of India's four major salt regions, namely Bengal and Northern India (the Punjab and the United Provinces), where the systems of administration contrasted sharply. Bengal was supplied from overseas and taxed through the customs, and the editorial soon dismissed its relevance: the danger of an invasion of foreign salt was now no longer real. Northern India, however, was another matter; it was supplied, in a manner reminiscent of Ho-tung, from the Sambhar salt lake in the Rajputana which was jointly owned by the Maharajas of Jaipur and Jodhpur, and it was administered through a system of taxation at source (the *YCTC*'s term was the familiar *chiu-ch'ang cheng-shui*) and free trade thereafter. Dane himself had been commissioner of salt revenue in Northern India, and the system he advocated in China was derived from the principles applied there.[82]

The editorial did not criticize these principles directly, but sought rather to turn the Indian example to its own advantage. It pointed out that the officials in India took a more active role in organizing production and transportation than Dane was allotting them in China. It argued too that circumstances in India and China, between which Dane saw "a close analogy," [83] were in fact different, particularly as regards transportation. In India, the taxation-at-source system worked because of "the advantage of the universality of railways." But "our country's communications are not advantageous, the railways are not universal . . . so transportation organisation cannot be abolished: *kuan-yün, shang-yün,* one of the two must remain." [84] Yet the tone of the editorial was not consistently hostile: its burden was that there was room for grafting onto Dane's system the reformers' favorite notions of government acquisition of salt plus officially supervised but privately conducted trade. The editorial concluded by stressing the common features of the two systems: taxation at source, limitation of production centers, uniformity of tax rates.

1914 thus saw the beginning of friendlier relations between Dane and the Chinese reformers. An editorial in the *YCTC* reviewing the achievements of the year asserted that the genuine reforms, such as standardization of tax rates and abolition of wastage allowances, were mainly due to Dane; while the failures, such as failures to establish proper institutions to regulate production or to replace artificially evaporated salt by solar-evaporated salt, were due to the Chinese authorities. The prospects for 1915, it concluded, gave grounds for hope.

In fact, 1915 produced increased difficulties for the reform with the return to office of Chou Hsüeh-hsi: the resistance of the salt merchants now had the support of the ministry of finance. This brought Dane and the reformers closer together. The August 1915 number of the *YCTC* carried a translation of an editorial from the *Far Eastern Times* urging Chang Hu and Dane to stand firm in their joint resistance to merchant privilege, and an editorial in the same number attacked the policy of raising benevolences from the salt merchants, which Chou Hsüeh-hsi had revived and which Dane rightly saw as subversive of the whole system established by the reorganization loan agreement.

Towards the end of 1915, Ching Pen-po had to abandon the salt agitation because of his differences with Yüan Shih-k'ai over the monarchy. He did not resume it again after Yüan's death, partly because he was preoccupied with the development of his own salt-refining company, but it is also possible that he no longer objected to the policy's being pursued. The incipient alliance of 1915 foreshadowed the gradual adoption of Dane's own program by the reformers in the 1920's, which reached a climax in the new salt law of 1931, the first article of which stated: "Salt shall be taxed at source (*yen chiu-ch'ang cheng-shui*): the people shall have freedom to buy and sell; and no one at all shall obtain a monopoly." [85]

Dane's relationship with the salt interest as a whole was thus a positive one: at each level — officials, merchants, and reformers — he found significant areas of support and could contribute towards achieving their desired objectives. Equally, the salt interest contributed to Dane's success. The officials contributed the minimum of cooperation he needed to initiate his system; the merchants contributed the economic dynamism without which it would not have been viable; and the reformers contributed a climate of opinion favorable to change and men in key places.[86] More than the Maritime Customs, at least in Hart's day, the reformed salt gabelle was a genuinely Sino-foreign undertaking. The *North China Herald* was no more than accurate in its comment: "There is no intention to claim omnipotence for the foreigner. Sir Richard Dane is the first to acknowledge that the results obtained in the Gabelle would have been impossible without the cooperation of the Chinese themselves. It is the Chinese themselves who have effected reorganisation. Sir Richard Dane only told them how to do it. They did it." [87]

7 | The Structure of the Salt Administration in 1920

Between 1900 and 1920, the traditional salt administration was subject to a variety of pressures, exerted both by the impersonal march of events and by the deliberate policies of administrative reformers. By 1920, it will be argued, the cumulative effects of railway construction, of the financial necessities of late Ch'ing and early republican government, and of the reforming efforts of the imperial governors-general, the revolutionaries of 1911, the Salt Society, and the foreign inspectorate had together brought about a significant degree of change in the entire structure of the salt complex. At every level, there is evidence of change: limited change in the field of technology; substantial change in the field of commerce; and total, that is structural, change in the fields of taxation and administration.

These changes were not traditional in character, nor were they simply haphazard. On the contrary, they fell into the dual pattern of modernization outlined in Chapter 4: stepped-up efficiency, diminished preponderance; the simultaneous increase in dynamism and decline in relative status which the units of an institutional complex undergo when the number in the complex increases.

On the one hand, the various components of the whole commercial-fiscal network became more versatile, better organized, and more effective through a variety of modernizing innovations: diversification of demand, increase in productive capacity, mechanization of transport, acceleration of turnover, centralization of financial control, and specialization of personnel. On the other hand, the units of the salt administration relinquished power to other institutions or to autonomous regulative mechanisms: the discretionary power of officials was curtailed by treaties and regulations, the foreign administrators exerted less influence in domestic politics than their Chinese predecessors, direct government operation of the trade was abolished except in Manchuria, private enterprise replaced merchant monopoly in half the inspectorate areas,[1] and official supervision of supply, markets, and prices was everywhere relaxed. The changes between 1900 and 1920 thus satisfactorily fulfill our criteria for modernization.

The process, however, was not carried through to completion: the salt administration had not developed some of the attributes it needed for greater sophistication, while at the same time it clung to powers which should have been relinquished. On the one hand, the demand pattern had not been fully diversified, production methods had not escaped from dependence on the rhythms of nature, and transportation was far from fully mechanized. But equally, arbitrary interference by officials was not yet eliminated, the restrictive practices of monopoly merchants still impeded the expansion of the trade, and salt was still an unnaturally conspicuous commodity, not yet eclipsed by the growth of other industries of equal importance to the revenue and to the consumer. In 1920, there were, therefore, significant areas of the salt administration in which modernization had not taken place.

That modernization was not completed was due to external circumstances rather than to the inadequacy of the salt reformers or the resistance of traditional elements within the salt administration. Full modernization of the salt administration required changes beyond it: the development of a chemical industry, the completion of the communications network, the growth of a less static business outlook, and the establishment of new methods of state finance and bureaucratic remuneration. These changes in turn were hardly possible without a new political framework, which should provide minimum conditions of law and order, an incorruptible administration, and an internal common market. In the absence of these, the developments which would have made possible the full modernization of the salt administration could not take place, and modernization in the salt administration, while genuine and substantial, remained limited. Change in the single institution could not be completed without change in the total environment.

The same four aspects — technological, commercial, fiscal, and administrative — which in Chapter 1 were used to demonstrate the traditional character of the unreformed gabelle, will be used as analytical apparatus in detailing the results of the changes that took place between 1900 and 1920. This procedure is legitimate, it is argued, since the qualities attributed to each aspect there can in fact be seen to be negative qualities in terms of our definition of modernization. Technology was earthbound; commerce was stagnant; fiscal administration was limited in range; bureaucratic organization was not rationalized; all these are deficiencies of power. Equally, there were defects in the abandonment of power: the impersonal regulation of the salt complex by the processes of the law and the forces of the market was not fully conceptualized, much less practiced; too much was controlled as well as too little. Where these qualities can be shown to have been reversed, modernization will have been successfully demonstrated.

Technology: Persisting Patterns and Incipient Changes

Before the reform, the salt industry in China could be described as earthbound, in that the imperatives it obeyed were physical rather than social. The demand for salt was limited by the requirements of human metabolism, since the market was overwhelmingly alimentary; at the same time, because the crucial factor in manufacture was evaporation, powered predominantly by the sun or combustible plants, the supply of salt was at all times limited by weather conditions. This pattern had not been radically disturbed by 1920. Changes of a modernizing kind had indeed occurred, both in demand and in supply, but they remained marginal, less significant in themselves than as prototypes of future developments.

The demand for salt. The criteria for successful modernization here would have been a simultaneous expansion and diversification in market outlets. The consumption of alimentary salt may be expected to rise with modernization, since in the traditional economy, as a result of restrictive practices and distributive bottlenecks, even everyday dietary requirements frequently go undersupplied. At the same time, the relative importance of alimentary salt would have declined as nonalimentary customers, the chemical industry for one, became major consumers. In the period 1900–1920, some diversification did take place in both alimentary and nonalimentary demand, but on a strictly limited scale.

In the alimentary field, the principal innovation was the growth of a market for refined salt: this was a change in quality rather than quantity, but a modernizing one, nevertheless, because of the greater consumer discrimination involved.

In Europe, refined salt, "salt upon salt" as it was called in England since the process involved the dissolution of crude salt in water and its reevaporation, had been produced since the fifteenth century, its first major production center being the Low Countries. In China, however, refined salt did not appear until the nineteenth century, when small quantities were imported into the treaty ports for European consumption. Chinese soon acquired European tastes for salt, and Ching Pen-po described the effect on consumer preference: "it is human nature 'to love the good and hate the evil' . . . people saw that foreign salt was reasonably priced and excellent in quality." [2] Refined salt of Chinese manufacture first figured in the inspectorate reports in 1917, when an import of 9,615 piculs from Tientsin was noted at Hankow. Thereafter the trade grew briskly, finding its main outlet in Hupei, Hunan, Kiangsi, and Anhwei.

Salt from Huai-nan, the regular source of supply for these four provinces, was unusually poor in quality. Produced by the expensive *chien* or boiling method, it had to compete against cheaper *shai* or solar-evaporated salt from Huai-pei and Kwangtung, and the only saving open to the Huai-nan merchant was adulteration. Ching Pen-po's associate Tso Shu-chen claimed that of the 6 million piculs of salt consumed annually in the area, over a million consisted of mud and grit.[3] It is not surprising, therefore, that a market for a superior product existed.

The supply of Szechwan salt which had formerly met this demand (Szechwan salt being regarded as equivalent in quality to foreign refined salt) was, in the period after 1916, interrupted by political conditions on the upper Yangtze. Table 19 shows that the arrival of Chinese refined salt in the Yangtze provinces coincided with a fall in imports from Szechwan and, it may be supposed, was in part a response to this fall. Refined salt, however, was not simply a substitute for Szechwan salt. The distribution pattern was significantly different; while Szechwan sales had been concentrated in Hupei and northwest Hunan, refined salt sold throughout the middle Yangtze. In 1922, Hupei took 206,522 piculs, Hunan 45,-360, Kiangsi 202,450, and Anhwei 32,079. A definite mutation in consumer taste must be postulated as another factor in the rising sales of refined salt.

Table 19. Imports of Szechwan and refined salt in Hupei, Hunan, Kiangsi, and Anhwei, 1916–1922 (piculs).

Year	Szechwan imports	Refined imports
1916	1,212,500	–
1917	1,084,034	9,615
1918	625,500	68,233
1919	631,800	116,358
1920	529,200	241,456
1921	837,000	–
1922	602,100	486,411

Source: District Reports.

This change in taste was a significant development: it was a small indication of the Europeanization of the Chinese consumer. Its quantitative importance, however, must not be exaggerated: the half-million piculs of refined salt consumed in the four Yangtze provinces constituted only 8 percent of the total demand in that area as late as 1922.

The same caution must be applied to a further development in the demand for salt, namely, the expansion of the nonalimentary market. Fishery salt, the chief nonalimentary component of the traditional demand in both Europe and China, led the way. Huaipei, for example, which supplied part of the salt needed for the fisheries of the Yangtze delta, increased its issue of fishery salt from 33,041 piculs in 1916 to 136,924 piculs in 1921. The quantity involved is still small.

The opening of an industrial market for salt was likewise more important for what it portended than for its quantitative impact. In 1921, Fan Hsü-tung, managing director of Ching Pen-po's Chiu-ta Salt Refining Company, founded the Pacific Alkali Company to pioneer in China the manufacture of soda by the Solvay ammonia-salt process. The factory was adjacent to the Chiu-ta works. Output was boldly planned at fifteen tons a day. After initial difficulties, technical and commercial, production expanded rapidly, and "by 1937, domestic production of soda-ash had reached between 180 and 200 tons a day and the volume of sales inside China was higher than the total of imports from Britain for that year." [4]

In 1921, when the figures for industrial salt were first recorded, the salt administration released only 61,000 piculs of industrial salt; by 1925, the figure was 148,000, and by 1933, it was over 2 million. The meaning of industrial salt was formally defined in article 6 of the new salt law of 1931: thirteen categories were listed, including soda, hydrochloric acid, sodium sulphate, potassium sulphate, dyes, soap, glass, and paper.

In quantitative terms, however, nonalimentary demand in the early 1920's, whether from the fisheries or from industry, was still only of marginal importance. In Manchuria in 1922, for example, out of a total quantity of 4.5 million piculs released for sale, nonalimentary salt, in this case fishery salt, amounted only to 21,922 piculs. Even in 1925, when the chemical industry had got underway, nonalimentary salt amounted only to 8 percent of the total salt issued in China.[5] Thus, alimentary predominance in the demand for salt had not been shaken by the changes between 1900 and the early 1920's.

The supply of salt. Salt production likewise remained largely traditional. Except for the thirteen refineries either in operation or projected by 1922, whose output amounted to only 1.5 percent of the total, no new processes were introduced.[6] The availability of the traditional types of fuel continued to be the controlling factor in production methods, and, except in Szechwan where natural gas was used, salt production remained tied to solar energy and the reed crop.

As a result, the production side of the industry retained its dependence on seasonal factors. In Huai-nan in 1920, for example, the fall in production was explained by "the short supply of fuel reeds in the first half of the year on account of the ravages of locusts, and in the latter half of the year the very unfavourable climatic conditions." In Ch'ang-lu in 1921, salt works were damaged by a sandstorm, while in Shantung, "The year was a very wet one and production in all parts of the District was smaller than usual." In 1922, the Huai-pei inspectors complained of "a great storm which destroyed a large portion of the salt stocks kept at the works, as well as the works themselves." Even the more sophisticated Szechwan industry could be affected by natural disaster: in 1917, for example, production was reduced by an outbreak of rinderpest among the buffaloes used for drawing the buckets from the wells.[7]

In one respect, indeed, the manufacture of salt actually retrogressed. In the period 1900–1920, *shai* or solar-evaporated salt increased its percentage of total output at the expense of *chien* or boiled salt. This development, which was related to transportation factors, will be discussed in more detail below. Here, it may be noted that, since *shai* production required less capital than *chien*, and since higher capitalization may be regarded as an index of technological modernization, the shift to *shai* was a backward move despite its rationality in terms of costs.

The only change in salt manufacture in the early republic which had any modernizing significance was an increase in productive capacity. According to tables given in the eighteenth issue of the *YCTC*, the average annual production of salt in China for the three years preceding the reform, that is for 1910–1912, was 33,-895,977 piculs. The period 1913–1916 saw Dane's reform come into operation, and in the next triennium, 1917–1919, the average annual production had risen to 43,687,725 piculs.[8] Thus, the years of Dane's reforms established new levels of productive capacity: the boom year 1919, with its production figure of 48,354,855 piculs for the whole of China, foreshadowed the quinquennial average for 1931–1935 (excluding Manchuria) of 46,302,000 piculs.[9]

However, the market response to this increase in production illustrates graphically the extent to which the continued predominance of alimentary demand restricted the supply pattern. Manufacturers, encouraged by Dane's lifting of restrictions and by the availability of new forms of transportation, had hoped for higher sales and raised productive levels from 33 to 43 million piculs, an increase of 30 percent. The market, however, still governed by its alimentary component, could not absorb anything like this level of increase. Consumption of salt has a built-in tendency to expand, given that population rises steadily and provided that the new consumers have money with which to buy. But a population increase of 2 percent per annum from 1912 onwards would, by 1919, have produced an increase in demand of a little under 15 percent, well below the amount being produced.

By 1920, the industry was facing a temporary crisis of overproduction. Already in 1918, the Hai-chou inspectors spoke of the "general over-production in Huaipei." In 1919, the Ch'ang-lu inspectors noted that "If the saltmakers are allowed to go on manufacturing salt without any restriction, it will be very difficult to arrange for the clearance of saltworks and the storage of salt at the depots in 1920," and the following year, Dane's successor, Sir Reginald Gamble, did introduce limitations on production. In the same year, the Fengtien inspectors too stated that "production of salt in this District is always in excess of requirements." [10]

In 1920, salt production was forced back to 36,436,367 piculs. So long as the demand pattern was dominated by alimentary salt, there was bound to be tension between heightened productive capacity and the limited absorption capacity of the market which frustrated it. This tension could be resolved only by the growth of a modern chemical industry requiring massive quantities of salt. The Pacific Alkali Company pointed the only way out of the impasse.

Commerce: The Railway Revolution

The commercial aspect of the gabelle, that is, the system of distribution, experienced appreciably more change between 1900 and 1920 than did the technological side. The impact of modernization here was both substantial and unmistakable: the distribution system no longer retarded an increase in the volume of sales.

In 1900, any expansion in the volume of salt trade had been held back by the marketing system. This system, as a result of being based on inland water routes, presented three obstacles to further development: the communications grid was inadequate, principally by reason of the defective north-south axis; growth potential in the commodities for which salt was exchanged was negligible and this trading weakness automatically limited the growth of salt sales; and third, because of the traditional means of transport and the bureaucratic red tape along the way, the turnover of goods was slow and irregular.

By 1920, railway construction had considerably ameliorated these conditions. The railway provided for the first time an adequate north-south axis; by enhancing the importance of the northeast salines, it promoted the growth of a more dynamic pattern of exchange; and, together with streamlined administrative methods, it had the effect of accelerating turnover. Where the railway had been introduced, there was no longer any facet of the marketing system which obstructed expansion of sales. The national grid, however, was still far from comprehensive, so that the overall modernization of the commercial side had to remain incomplete, just as technology had fallen short of complete reform.

The pattern of routes. In 1900, China possessed in the Yangtze an uninterrupted artery of east-west communication. The north-south axis, on the other hand, that complex of waterways which ultimately linked the Pearl River and the Amur, suffered from four major defects. First, it was discontinuous, portages being necessary between the Sungari and the Liao in Manchuria, and between the Kan and Pei rivers in south China. Second, it was seasonal: between the beginning of December and the end of March, Hosie had noted in 1900, "all water-borne traffic ceases to the north of Newchwang," and the northern sections of the Grand Canal were likewise frozen over in winter. Third, it might on occasion be unusable: the Grand Canal frequently lacked water, the Tientsin customs commissioner reporting in 1911 that the stretch in Shantung beyond Lin-ch'ing was "dry and useless." Finally, even when open, the route suffered from congestion: it took on the average three or four days to cover the two miles through the northern suburbs of Tientsin, and might even take as long as thirty.[11]

By 1920, the railway had transformed the route. There was now a continuous rail line from Harbin to Changsha and, after a gap of 270 miles, from Shiuchow to Canton. The movement of goods was no longer seasonal. In 1921, the Newchwang customs commissioner noted that the "facilities offered by the South Manchuria Railway of economical transport to ready markets at all seasons . . . have made it impossible for junks to compete with it," and that the number of junks operating had thus been reduced from

20,000 to 4,000. The railway, furthermore, was immune from water shortage, silting, and sandbars; the worst obstruction reported was an occasional washout due to heavy rain, as for example on the Peking–Hankow line in the summer of 1919. Congestion was no longer chronic: "abnormal" was the description given in 1918 by the Kirin–Heilungkiang auditors when, as a result of the Siberian expedition, "Both the South Manchurian and Chinese Eastern Railways were congested by the transport of troops and military stores." [12] Except for the awkward gap in southern Hunan, which was closed in 1935, the north-south communications were now no less convenient than those between east and west.

Those salt divisions which lay along the north-south axis took full advantage of the new facilities for the transport of salt. A writer in the *YCTC* described the situation in Ch'ang-lu: "Upon the completion of the Peking–Mukden, Peking–Hankow, Peking–Kalgan and Tientsin–Pukow lines, salt was increasingly transported by rail, and for convenience of rail transportation, Ch'ang-lu had the edge over all other salines." Dane's report noted that at Han-ku, the principal Ch'ang-lu saline, salt transported from the works by rail increased from next to nothing in 1913 to nearly 60 percent in 1917. In Manchuria, too, the railways were extensively utilized: in 1915, "salt destined for Kirin was, for the most part, transported by the South Manchurian Railway," while "Salt destined for Heilungkiang was shipped by sea to Vladivostock and then by the Chinese Eastern Railway to the Central Depot at Harbin," a procedure changed in 1917 to direct rail shipment to Harbin by the South Manchurian railway. Similarly, in the south, a casual comment of the Kwangtung inspectors in 1920 reveals the salt trade's dependence on the railway in the key northern section of that division: "The Canton–Hankow Railway track was destroyed in several places and this caused the transportation of salt to be temporarily suspended." [13]

In all these areas, the salt trade expanded with the conversion to rail transport. In Ch'ang-lu, sales expanded from 4 million piculs in 1900 to 6 million in 1922; in Manchuria, over the same period, sales rose from 3.6 million to 4.5 million; and if in Kwangtung the increase was only from 4 to 4.5 million piculs, it must be remembered that the Canton–Hankow line was incomplete, and the division could not exploit its natural market in southern Hunan.[14]

The pattern of exchange. In a traditional economy, the variety of exchanges possible is largely a function of the communications available: commerce is shaped by the routes it follows. In China in 1900, the exchange pattern for salt, produced by the reliance on inland waterways, was relatively simple. In most cases, salt was marketed as an urban export to acquire basic foodstuffs or raw materials, such as grain or timber, as for example in the Canton–Kwangsi exchange of salt for firewood; or it was sold as a rural export in return for basic consumer goods, particularly textiles, the paradigm here being the Szechwan export of salt to Hupei to finance the import of foreign cottons from Hankow. Neither linkage offered much growth potential: urban demand for grain and timber grew slowly,[15] and while a rural area might want more manufactured goods, it was difficult to expand sales of alimentary salt sufficiently to pay for them.

The period 1900 to 1920 saw the emergence of a more complex and more dynamic exchange context. This resulted from the rise of a single, nationally dominant region for salt production in northeast China, based on the improved north-south axis and also on the low cost of *shai* or solar-evaporated salt.

Table 20. Regional percentages of salt production in China, 1900–1935.

Region	1900	1919	1931–1935
Northeast	35	58	57
Southeast	41	21	23
Interior	24	21	20

Source: Customs Report; District Reports, 1919–1921; Tseng Yang-feng, *Chung-kuo yen-cheng shih.* The northeast covers Manchuria, Ch'ang-lu, Shantung, and Huai-pei; the southeast, Huai-nan, Sung-kiang, Liang-che, Fukien, and Kwangtung; and the interior, Ho-tung, the northwest salines (K'ou-pei, Chin-pei, Hua-ting), Szechwan, and Yunnan. Tseng's average figures for 1931–1935 do not include any entry for Manchuria; a low figure of 3,698,000 piculs has therefore been assumed to give a total Chinese production in those years of 50,000,000 piculs.

Tables 20 and 21 summarize the evidence for this shift. In 1900, salt production had been evenly distributed along the coastline of China; by 1919, a pattern of northeast dominance had emerged; and by the 1930's, it was well established. In the early republican period, the change is revealed by the production figures for Ch'ang-lu, Huai-pei, and Huai-nan. Ch'ang-lu and Huai-pei followed a common pattern of growth to a high point in 1919, followed by a fall; this fall, however, was to a plateau substantially above the prereform figure, a level which foreshadowed those of 1931–1935. Huai-nan, on the other hand, declined steadily and by the 1920's was ceasing to exist as an independent source of salt. In 1922, of the 4,763,382 piculs of salt issued at Shih-erh-wei for consumption in Huai-nan, only 334,235 originated in the division: all the rest was Huai-pei salt.

Table 21. Salt production in Ch'ang-lu, Huai-pei, and Huai-nan, 1900–1935 (piculs).

Year	Ch'ang-lu	Huai-pei	Huai-nan
1900	4,000,000	1,400,000	5,000,000
1910	4,538,648	1,461,667	3,628,053
1911	3,249,068	675,207	1,890,761
1912	3,567,552	2,290,157	2,411,445
1913	7,018,094	3,968,265	2,882,617
1914	7,113,906	5,584,430	2,093,370
1915	3,605,286	3,542,568	2,236,085
1916	7,576,130	4,108,561	1,978,580
1917	9,847,547	8,337,302	3,198,687
1918	6,067,789	6,660,210	2,994,559
1919	11,899,996	11,221,269	2,863,536
1920	6,244,427	5,405,449	1,404,157
1921	6,691,085	4,908,578	1,057,185
1922	5,837,576	5,271,711	554,196
1931–1935 (average)	6,816,000	8,495,000	1,191,000

Source: Customs Report; *YCTC*, No. 18; District Reports; Tseng Yang-feng, *Chung-kuo yen-cheng shih.*

The rise of the northeast salines and the reorientation of routes associated with it are clearly seen in Honan. In 1911, the province was divided between four salt divisions: Ch'ang-lu (fifty-two districts), Ho-tung (thirty-four), Huai-pei (fourteen), and Shantung (nine). This was a well-established distribution of markets: it corresponded to the traditional system of rivers and roads described by Richthofen in 1870, while the Yung-cheng editions of the Ch'ang-lu and Ho-tung salt laws show respectively fifty-six and thirty-one Honan districts belonging to their divisions.[16] By 1920,

Ch'ang-lu had entirely captured the fourteen Huai-pei districts "by its advantage of railway transportation" and was legally permitted to compete in eight of the Ho-tung districts. The effect on Ho-tung, indeed, was more serious than this figure might suggest. In 1919, the Ho-tung inspectors noted that "the encroachment by Changlu salt sellers into the Hotung eastern area, made possible by the cheaper railway haulage," was "a very serious matter, and it is affecting all trade in the south-western part of this Province." [17] Honan, initially divided between four salt divisions by various river and road systems, was becoming a satellite of Ch'ang-lu via the Peking railway system.

This development of regional specialization based on low costs and transportation advantages can be paralleled in other parts of the world and may indeed be regarded as the typical line of advance for the salt trade in early modernization. In the fifteenth-century western Mediterranean, for example, Jacques Heers notes "une concentration assez nette de la production en quelques lieux privilégiés." [18] The smaller producers such as Cagliari, Hyères, and Roussillon gave way before a few major centers: Iviça, La Mata, Trapani. In nineteenth-century France, the once dominant Biscayan salt retreated before its rival in the east, well salt from Lorraine which enjoyed superior railway facilities. In early twentieth-century India, the supply of salt for Bengal, Bihar, Assam, and Burma was dominated by the Red Sea area. Like northeast China, Aden enjoyed exceptional advantages in production and distribution. Minimal rainfall created an ideal terrain for solar evaporation once the proper methods were introduced in 1886 by an energetic Italian from Trapani, Agostino Burgarella, while its situation on an international trade route allowed minimal freight rates.

In China, the rise of the northeast salines altered the exchange pattern of the salt trade by throwing into relief the new and more complex commercial connections characteristic of this developing area. In 1900, the salt trade in both Manchuria and Ch'ang-lu had followed the traditional salt-grain pattern; by 1920, new exchanges had become dominant. In Manchuria, the soybean industry, whose output rose from 600,000 tons of beans in 1900 to 4,200,000 tons in the early 1930's, came to be regarded as the staple. As the Dairen customs commissioner put it in 1921, "Economic conditions in Manchuria hinge entirely on the state of the bean, beancake, and bean oil market." The quantity of salt going north was therefore closely related to the value of bean products coming south. Thus in 1920, the Fengtien inspectors noted that "The sudden panic last spring and the subsequent state of commercial depression which prevailed in the Manchurian markets during the year, resulting in the gradual fall of prices of Manchurian staple products, affected considerably the financial position of the inhabitants, who had to observe strict economy in consumption of daily necessaries, including salt." [19]

Bean products, however, were neither a traditional foodstuff nor a basic raw material to be consumed in the South Manchurian cities. They were industrial raw materials, prototype industrial oil and artificial fertilizer, which were reexported from South Manchuria to the ports of south China, to Japan, or even to Europe: beancake, for example, was sent to Amoy and Swatow as fertilizer for the sugar plantations, whose crop was then sold in the Yangtze valley in exchange for rice. A similarly complex pattern of exchange was also developing in Ch'ang-lu between salt going south and raw cotton supplies from Chihli coming north to be reexported to Japan or processed by the new Chinese machine textile industry. Thus, in the northeast, a multilateral pattern of trade was replacing the simple town-country exchange of the traditional economy, and the increased importance of the northeast made this change of national significance.[20]

The new exchange pattern imposed fewer restrictions on the growth of the salt trade than the old. As the Newchwang customs commissioner stated in 1911, "the bean industry . . . seems capable of infinite expansion in the future." [21] The market in Manchuria for alimentary salt had, no doubt, its saturation point, but at least the salt trade would not be prevented from reaching this point by a sluggish trading partner. Similarly, raw cotton for modern-style mills had an unlimited expansion potential, particularly in Chihli, which, over the next two decades, rose to be China's leading producer of raw cotton. Affiliated now with a more dynamic product, it is not surprising that the consumption of salt in Chihli rose from 2,800,000 piculs in 1900 to 4,700,000 in the early 1930's.

The turnover factor. The traditional cycle of the salt trade had moved slowly. In Ho-tung, the time limits for the transportation of salt were reckoned on a basis of ten miles a day for oxcarts and seventeen for donkeys, mules, and boats; even these speeds were not always to be relied on, the journey to Sian, for example, a distance of 200 miles, being given a time limit of twenty-four days.[22] In Szechwan, the journey from Tzu-liu-ching to the mainstream of the Lu River took over a month, though a train, it was calculated, could do the same distance in two to three hours. Wolf was right therefore when he spoke in 1901 of "the clumsy and tedious mode of transport" as one of the heaviest burdens of the trade.[23]

By 1920, modern communications were beginning to accelerate this cycle, making it faster and more regular, particularly in the two northeastern divisions of Huai-pei and Ch'ang-lu. In 1921, in response to their enquiry "whether the merchants prefer direct shipment from the works to consumption areas by steamer or the present means of junk transportation with Shiherhwei as transhipment centre," the Huai-pei inspectors found that "the merchants are in great favour of direct shipment of salt, firstly because of the quicker transportation and consequently less capital held up, and secondly because there is less danger of wastage or wreck." [24] This is not surprising, since the steamer could cut the journey up river to less than a week. Similarly, in Ch'ang-lu, the railway could deliver parcels from Peking to Hankow in two days, whereas the traditional horse-post time had been fourteen. The faster turnover meant less tying-up of capital; the rate of profit increased, providing an incentive to the entrepreneur to lift the volume of salt transported.

The impact of quicker transport was amplified by administrative streamlining. In Ch'ang-lu, where before the reform there was much red tape, the "Yin Piao and miscellaneous Passes were abolished and a single form of Salt Transportation Pass was introduced"; separate checking stations were eliminated; and duplicate receipts for tax, involving only the merchant, the commissioner, and the inspector, replaced receipts in quadruplicate. In 1918, Ch'ang-lu experienced a temporary shortage of rolling stock, but, "Fortunately, the procedure of packing salt in the depôts had been well established, so that 10 empty cars shunted into the depôt could be fully loaded with salt within two hours," and supplies could be maintained by rapid turnover. In Huai-nan too, procedure was simplified. In 1920, a system of direct, individual payment to the bank replaced payment through the merchant guilds: "This helped the merchants to handle their business easily and quickly. Since that time salt junks are being cleared from Shiherhwei soon after payment of duties is effected." [25]

Faster turnover, however, like the other benefits of the railway, tended to peter out beyond the trunk lines. In 1916, for example, the Kirin–Heilungkiang auditors complained of the amount of capital locked up in salt depots remote from the railway. The Hankow customs commissioner wrote in 1911: "Much as railways will develop the wealth of the province, only a small fraction of the gain can be achieved until the roads to feed them exist." [26]

The situation in Hupei in 1921 illustrates this lack of secondary routes.[27] In that year, tax in Hupei was made payable on import into the province, instead of when the salt was sold to the local dealer. As a concession, however, payment was allowed in promissory notes instead of cash, the date of maturing varying according to the time by which the merchant might expect to acquire the cash to meet his check. In the case of salt sold in the Hankow region, the notes matured within 7 days, but in the case of upcountry destinations, maturing dates of between 25 and 150 days were set. It is not suggested that these dates represented normal traveling times to and from the place of sale: delays in payment by local dealers and difficulties of exchange would also enter into them; but they do suggest that beyond the railway, old rhythms still prevailed. In commerce, as in technology, modernization was incomplete.

The Fiscal Aspect: Assessment Comprehensive, Receipts Consolidated

At all periods in its history, the salt administration was primarily a fiscal machine, so the degree of modernization in this aspect achieved by 1920 is of particular importance in establishing the overall level of modernity in the organization.

A tax-levying agency may be regarded as fully modern if it universalizes the assessment of the object liable to tax, and if it channels all receipts into a consolidated fund at the disposal of the taxing authority. The agency's power must be maximized in terms of both extension and concentration.[28]

Already in 1916, the salt administration effectively fulfilled these two criteria: all salt was being subjected to tax and all tax was being paid into a single account. This achievement, furthermore, was not annulled by the period of warlord anarchy after 1916: the inspectorate continued to polish up its assessment system and was able to go on channeling all receipts into the consolidated account in the collecting banks, even though subsequent transfer to the salt revenue account with the group banks at Shanghai might be impeded. As regards its fiscal aspect, it is not too much to say that modernization was by 1920 complete.

Assessment of tax. In 1900, assessment had been circumscribed by the limited administrative range of the traditional bureaucracy and hence had been inequitable in terms of taxable capacity: rich, urbanized regions close to the salines were underassessed; while poorer, rural areas, more vulnerable to taxation because of their dependence on easily controlled transportation routes, carried an excessive load. In Chapter 1, Szechwan division and the Yangtze delta districts of Huai-nan and Liang-che were shown to be conspicuous examples of this situation; these areas will be used again to illustrate what had been accomplished by 1918, Dane's last year in office.

In Szechwan salt division in 1900 (see Table 2 in Chapter 1), Szechwan province proper, that is, the *p'iao-an* and *chi-an,* had been seriously underassessed, while the export trade to Hupei, Kweichow, and parts of Yunnan had been heavily mulcted. Table 22 shows that, by 1918, these defects had been remedied. Within Szechwan proper, it is true, *p'iao* or land-carried salt was still less heavily taxed than *yin* or waterborne salt, and beyond Szechwan, the *chi-ch'u* or export salt for Hupei still bore a disproportionate share of the tax. This, however, could be justified as protection for Huai salt in Hupei, which, anyway, carried a heavier tax load than Szechwan salt, but the major inequalities, both of tax base and of rates of tax, had been eliminated. More *p'iao* salt was being subject to tax, while the inequality of rate between the *chi-an* and the *pien-an* had been removed.

Similar inequalities had existed in the Yangtze valley in 1900.

Not only did the Yangtze delta districts of Huai-nan and Liang-che pay tax at a much lower rate than the up-river provinces, but their quota of salt (which alone paid tax) seriously underestimated the true consumption, while the up-river quotas were reasonably realistic. Dane was less successful here in equalizing tax rates than he had been in Szechwan. Though rates in the *shih-an*, that is, in the Kiangsu districts of Huai-nan, were raised from between $0.60 and $1.46 a picul to between $0.75 and $2.00, these remained well below the $4.50 levied on salt going to Anhwei, Kiangsi, Hupei, and Hunan. In the Kiangsu districts south of the river formerly belonging to Liang-che and erected by Dane into the separate division of Sung-kiang, more was achieved, rates being raised from between $0.240 and $2.376 to between $1.00 and $3.20.[29]

Table 22. Salt releases and revenue in Szechwan in 1918.

Area	Salt releases (piculs)	Percentage of total (1900 percentage in parentheses)		Revenue (Chinese dollars)	Percentage of total (1900 percentage in parentheses)	
P'iao-an	2,674,377	45	(25)	3,699,031	31	(12)
Chi-an	1,775,600	30	(33)	3,962,422	33	(14)
Chi-ch'u	625,500	10	(18)	2,019,230	17	(34)
Pien-an	932,600	15	(24)	2,250,914	19	(40)
Szechwan	6,008,077	100	(100)	11,931,597	100	(100)

Source: District Reports, 1918. The *chi-ch'u* tax figure includes the levies made at Ichang and Shasi.

Dane, however, did succeed in bringing a larger percentage of the salt consumed in the delta under taxation. Before the reform, the quotas had been fixed at 381,360 piculs for the Huai-nan *shih-an* and 408,720 piculs for the Kiangsu districts of Liang-che; in 1918, the figures for salt taxed in these two areas were 1,058,490 and 875,708 piculs respectively, far more realistic figures for this populous area.[30]

In other areas also where the tax levied was out of line with salt consumed or traded, notably the Shantung peninsula and along the land route leading from Pakhoi into Kwangsi, the assessment was made more equitable and comprehensive. By the early

1920's, there was no major source of salt and no major salt route which was not making a substantial contribution to the republic's revenue.

The receipt of tax. Dane's centralization of the salt tax receipts has already been described in Chapter 4 and this achievement needs only to be summarized here. In 1900 only part of the gross collection of salt taxes was reported to the throne by the provincial authorities: items of taxation were left out of the reports, and the sums collected under each item were not always stated in full. Of the amount reported to Peking, only part was genuinely at its disposal, the rest being already earmarked for particular, provincial items of expenditure and not easily diverted to other purposes. Of the four main categories of salt tax — salt tax proper (*cheng-k'o*), likin, price increases (*chia-chia*), miscellaneous taxes (*tsa-k'uan*) — only the first, which in 1900 was worth 11 million dollars, was generally unencumbered.[31]

By 1920, on the other hand, the salt revenue available to the central government amounted to 55 million dollars, and, in the entire period 1914–1922, it never fell below 54 million dollars (see Table 12 in Chapter 4). Chinese as well as Europeans recognized that a radical change had taken place. Tso Shu-chen, for example, one of Ching Pen-po's circle, wrote in 1922: "salt likin used to be left to the provinces and was not forwarded to the board; today the salt taxes have all reverted to the central authorities and the situation is utterly unlike previous circumstances."[32] All the various salt taxes had been merged into a single item at the disposal of the central government.

The transfer of control from the provinces to the center was the point at which the reform of the gabelle made its greatest contribution to the politics of the early republic, in particular the success of Yüan's regime. In 1918, Jordan gave a favorable picture of Yüan's rule: "Before his death he had brought the provinces into a state of order and the writ of the Central Government ran from Peking to the borders of Tibet on the West and to Canton in the South." If Yüan momentarily achieved a new and more centralized government in China, it was partly as a result of the concurrent centralization in the gabelle, which in 1915 provided him with his principal source of revenue. The gabelle promoted centralization in other ways too: unlike the Maritime Customs, it kept its accounts in dollars rather than taels, a reform which Sir Charles Addis, the London manager of the Hongkong and Shanghai Bank, regarded as a greater service to China than the salt administration itself, since it accelerated the adoption of the dollar as the standard unit of account.[33]

It might have been expected that the rise of the warlords after 1916 would have seriously disrupted the work of the inspectorate. With its agencies deep in the hinterland of China, the salt administration was peculiarly sensitive to their activities. Interference with the salt administration took a variety of forms: refusal to permit the establishment of the inspectorate; insistence that revenue be paid to the warlord instead of to the central government; forced loans from salt merchants; looting of banks and intimidation of inspectorate personnel; and the taxing of salt illegally, outside the channels of the gabelle.[34] For example, at Swatow in 1918, the deputy commissioner, appointed by the Kwangtung tuchun Mo Jung-hsin, attempted to reintroduce merchant monopoly as a means of milking the trade, and the inspectorate in protest ceased to operate from January 1919 to November 1921. In Kwangtung itself in 1920, where the tuchun illegally extracted $4,750,000 from the salt merchants and a rival militarist imprisoned the Hai-lu-feng collector, "The District Inspectorate was like a leaking vessel struggling in a very rough sea." In Szechwan in 1921, the Ch'uan-nan inspectors went so far as to complain: "we have returned to the old regime, and are working with men who know nothing about salt affairs." [35]

To speak of a return to the ancien régime was, nevertheless, an exaggeration. Dane found means of minimizing the damage done by warlordism, so that while the work of the inspectorate might be temporarily disrupted by political chaos, the organization itself remained intact. The damage done was functional rather than organic, so that with the return of political order in 1928 operations could readily be resumed. In addition, the problem of warlordism for the salt administration was largely confined to the southwest provinces: the two most successful inspectorate divisions, Ch'ang-lu and Huai-pei, were not affected by it at all, nor was the Yangtze delta region.

Dane's initial countermove to warlord interference was subsidy: the warlord was paid a monthly allowance out of the salt revenue to induce him to permit the operation of the inspectorate, instead of attempting to tax salt himself. Dane would have liked the subsidy to have been a percentage of the revenue collected, so as to give the warlord a direct interest in protecting the inspectorate, but in fact, in the two most important cases where the system was applied, Yunnan in the period 1914–1917 and Kwangtung in the period 1916–1918, the subsidy was given in the form of a fixed monthly amount.

This technique for assuring minimum collaboration was extended in 1917 when the warlords began to demand the whole of the salt revenue collected in the areas under their control. The demand was accepted, provided the warlord would protect the gabelle and allow it to function normally in all other respects. The technique operated as follows: all salt revenue collected in a locality was made available to the local warlord, and no funds were in fact forwarded to Shanghai. However, to preserve the fiction that all funds were transferred from the local collecting banks to the group banks in Shanghai, Dane persuaded the consortium in August 1917 to agree to pay an amount equivalent to the warlord's withdrawal out of the reserve into the Shanghai current account. This amount was then repaid by the central government out of the salt surplus, the sums released to it by the group banks, into the Shanghai account.[36] The salt administration accounts registered the warlord's action as a "local withdrawal," and the only loser, of course, was the central government, which had to ratify what it could not prevent. Appearances, however, were thus saved and something of substance as well: for the framework of the inspectorate was maintained, and it required only the reopening of the pipeline from the local collecting bank to the current account at Shanghai to set the system once more genuinely in motion.

These, then, in outline were the techniques which allowed Dane, and after him Gamble, to preserve the integrity of their organization despite the diversion of revenue. They are seen in action most clearly in the principal area affected by warlordism, namely the southwest.

Southwestern warlordism evolved through three stages of increasing virulence. It began with Yüan Shih-k'ai's inability to impose his authority on Kwangsi and Yunnan, backward border provinces beyond the range of Yüan's railway-based troops and easily dominated by their own large military garrisons. In Kwangsi, Dane never succeeded in establishing the inspectorate at all. In Yunnan, too, at the outset of reform, "The Provincial Authorities parted with the collection of revenue with much reluctance," and although an inspectorate was set up in the autumn of 1913, it was not allowed to function properly throughout 1914. In January 1915, however, Dane himself visited the province and persuaded T'ang Chi-yao to accept a subsidy arrangement of $125,000 a month out of an average monthly revenue of $250,000.[37] The arrangement proved satisfactory, and in 1916 the inspectors commented: "the Yunnan Governor treated the District Inspectorate very fairly and courteously. The regulations were respected and the independent local authorities did not interfere with the work." [38]

The extension of southwestern warlordism beyond its original reservoir in Yunnan and Kwangsi began with the revolt against Yüan Shih-k'ai in 1916, which carried the influence of the southwestern tuchuns into Szechwan and Kwangtung. In Szechwan, the revenue for the four months, August to November 1916, was assigned to the Yunnanese tuchun Lo P'ei-chin, while in Kwangtung, the central government had to agree to pay the Kwangsi tuchun Lu Jung-t'ing "a sum of $2,000,000 in monthly instalments of $40,000" out of the salt revenue. Until the middle of 1917, however, thanks to the strong government of Tuan Ch'i-jui, interference remained limited: in Szechwan, the central government again received the revenues from December 1916 to May 1917, and in both provinces the salt commissioner remained its appointee.[39]

In the middle of 1917, the position worsened with the revolts against the central government which followed Chang Hsün's abortive coup. In Szechwan, from June onwards, all the revenue was commandeered by the local military commanders of either the central government, the Szechwan autonomists, or the southwestern forces, and by the beginning of the following year "the revenue of this important Province was entirely lost to the Chinese Government." In Kwangtung, in July 1917, the tuchun demanded a loan of $2,500,000 from the salt revenue, in January 1918 his subsidy was increased to $90,000 a month, while in March, "Dr.

Sun Yat Sen, as Head of the Military Government at Canton, prohibited the Bank of China from transferring any more money to the Salt Revenue Account with the Group Banks." [40]

Until 1920, however, the southwest confederacy headed by T'ang Chi-yao and Lu Jung-t'ing, which dominated Yunnan, Kwangsi, Kweichow, Szechwan, and Kwangtung from 1918, preserved a certain coherence, and the inspectorate could serve it as though it were another government. This was particularly the case in Szechwan, where, in April 1918, an agreement was signed with General Hsiung K'o-wu; this provided that in return for the grant of the whole of the Ch'uan-nan revenue, he was to protect the salt administration and prevent depredations by lesser militarists. Dane ratified this agreement on June 5 and the inspectors felt it worked well: "General Hsiung did all he could, in the circumstances, to protect the administration, salt manufacturers and transporting merchants and from the 3rd April the prospect of any direct or damaging interference with the administration, which, previously, had been a subject of great anxiety, was removed for the time being." [41]

The third phase of warlordism, beginning in 1920, was characterized by the disintegration of the large satrapies and the emergence of smaller units, whose chiefs, not controlling whole salt divisions, could not work through the gabelle and, instead, imposed taxes of their own on the salt trade. In the Pakhoi subdistrict of Kwangtung in 1921, the assistant inspectors noted: "The political strife caused us much embarrassment. For whenever the troops of either faction came here their Generals always demanded money." [42] In Szechwan, the position was as bad: no one militarist had control, and the inspectors noted eight illegal taxes being levied by various commanders. Yet even in post-1920 circumstances, some warlords in the area continued to work through the inspectorate. Under Ch'en Chiung-ming's rule in Kwangtung, the Swatow assistant inspectorate was reestablished after a lapse of nearly three years,[43] and in 1922 the area had a record year for salt brought under taxation. In Szechwan too, the machinery of the inspectorate continued to function and was even improved: in March 1921, an auditorate was established at Chungking to supervise the deposit of tax in the Bank of China, and, from August 1922, it was able to insist that payment be made in silver.[44]

In Yunnan, the inspectors increased their control of salt mining in 1921–1922 and were moving towards a partial nationalization of the industry. Thus, even at the height of warlordism and in the heart of warlord country, the inspectorate is to be found in operation, interrupted perhaps but never permanently halted. Fiscal modernization had come to stay.

Administration: A Specialized Service, Politically Impartial

As a department, the salt administration in 1900 was not distinguished from the rest of the imperial civil service: it was financed by the same premodern system of invisible taxation; its posts, many of which were sinecures, formed part of a single *cursus honorum;* and its members shared the undifferentiated politico-bureaucratic role of the traditional mandarin. The modernization of the department, that is, its rationalization, would consist in the reversal of these three trends.

By 1920, this had been largely achieved. The primary function of the old officials, the collection of tax, had been taken over by the inspectorate; this was a modern corps of administrators from which the traditional corrupt practices had been eliminated, and which offered a specialized career in a national agency to its members while, at the same time, it was detached from Chinese domestic politics. In addition, although Dane had no formal power to make the administrative branch conform to the standards of the inspectorate, the two services were becoming increasingly assimilated in personnel and orientation.

Corruption. The best available index of official "corruption" in the salt administration is the percentage of revenue collected which had to be expended on collection costs, that is, the difference between what was taken from the taxpayer and what the government received, expressed as a percentage of the former figure. Table 23 shows that by as early as 1916, the inspectorate had succeeded in reducing the ratio of collection costs to revenue collected in all areas except Shantung and Kwangtung. In Shantung, the licit salt trade was temporarily depressed thanks to political disorder and to smuggling from Kiaochow, while in Kwangtung, the increase in collection costs was due to the establishment in

1917 of the new assistant inspectorates at Pakhoi and Swatow: the previous year, costs had been only 6.31 percent of revenue collected. In overall terms, total department expenses fell from 17.01 percent of the revenue in 1913 to 11.37 percent in 1919. Ch'ang-lu and Huai-pei, the two most successful inspectorate areas, had, as may be noted from Table 23, particularly low ratios of costs to income.

Table 23. Comparison of salt administration costs in 1916 with those in the first year of the inspectorate.

Division	Cost as a percentage of revenue collected in first inspectorate year	Cost as a percentage of revenue collected in 1916
Ch'ang-lu	9.34	6.23
Fengtien	33.68	8.51
Shantung	7.43	9.29
Ho-tung	26.82	8.97
Liang-huai	47.36	–
Huai-nan	–	15.13
Huai-pei	–	4.404
Liang-che	14.59	12.73
Sung-kiang	–	7.38
Fukien	–	7.87
Kwangtung	7.30	13.94
Szechwan	10.71	–
Ch'uan-nan	–	5.64
Ch'uan-pei	–	9.82
Yunnan	21.83	11.67

Source: District Reports, 1913–1917. By 1916, Liang-huai had been divided into Huai-nan and Huai-pei, Sung-kiang had been separated from Liang-che, and Szechwan had been divided into Ch'uan-nan and Ch'uan-pei. The District Reports do not contain an expense percentage for Fukien in the first year of its inspectorate.

Specialization. It was argued in Chapter 1 from an examination of promotions recorded in the *shih-lu* (Table 7) that the salt administration in 1900 did not constitute a specialist service. A posting in it might involve specialist activities and knowledge of a particular branch of administrative law, but it was only an episode in the career of a theoretically omnicompetent official, who, in the course of his official life, might undertake other specialist activities as well: border affairs, copper administration, or Yellow River conservancy. Kuo Sung-tao, for example, besides serving a term as Liang-huai salt controller, was also called upon to act as naval expert and grain intendant, in addition to his better-known role as a diplomat.

That the new salt inspectorate, which was functionally equivalent to the salt commissioners of 1900–1912, was differently structured is made plain by the promotions of Chinese officials recorded in the single year 1922 (Table 24): in that year no one entered

Table 24. Promotion of Chinese inspectorate officials in 1922.[a]

Original appointment	Name	Subsequent appointment
D.I., Ch'ang-lu	Chen Shih-hsing	Not known
Auditor, Hankow	Tseng Kwang-chuan	D.I., Ch'ang-lu
D.I., Fengtien	Fung Ju-liang	D.I., Yangchow
Auditor, Ki-hei	Liu Nan	D.I., Fengtien
Secretary, Shantung	Li Chih-fan	A.D.I., Ch'ing-k'ou
Inspectorate, Chin-pei	Sun Kuan-ming	Secretary, Shantung
Inspectorate, Ningpo	Wang Chuan	A.D.I., Yang-chia-k'ou
A.D.I., Chefoo	Liu Tsung-yi	D.I., Huai-pei
Inspectorate, Yunnan	Li Yi-yu	A.D.I., Chefoo
A.D.I., Tung-t'ai	Chien Fang-tu	D.I., Ho-tung
Investigator, Yangchow	Chang Cheng-yu	A.D.I., Tung-t'ai
D.I., Sung-kiang	Yen Chia-shao	Auditor, Hankow
D.I., Huai-pei	Tseng Yang-feng	D.I., Sung-kiang
A.D.I., Swatow	Tang Shih-wan	Auditor, Ki-hei
D.I., Yangchow	Miao Chiu-chieh	D.I., Ch'uan-nan

Source: District Reports, 1922.

[a] "D.I." signifies district inspector; "A.D.I." signifies assistant district inspector. Romanization of personal names is taken from the District Reports.

the gabelle from outside, and no one left it for another branch of government service. In addition, the table indicates the extent to which the inspectorate had overthrown regionalism: compared to Table 7 in Chapter 1, there are few promotions within a locale. Even in 1922, in the middle of the warlord period, the inspectorate remained a national institution, able to make transfers from one end of the country to another.

Political involvement. In 1900, the salt administration was deeply involved in the Chinese political scene; the inspectorate, on the other hand, was insulated from it by its partly foreign character.

The foreign gabelle formed part of the treaty system. It had originated as additional security for a loan, and Dane and his colleagues, though employed by the Chinese government, enjoyed the privileges of extraterritoriality and were widely regarded as the instruments of the consortium. The original inspectorates were allocated with due regard for spheres of influence: a German was appointed to Shantung, a Frenchman to Yunnan, and Japanese to Manchuria and Ch'ang-lu. In particular, the foreign gabelle reinforced British predominance on the China coast. Of the forty-one foreign members of the gabelle in 1917, fifteen were British, no other power having more than six. Dane soon became, with Aglen and G. E. Morrison (formerly *Times* correspondent in Peking and at that time, 1912–1916, political adviser to Yüan Shih-k'ai), one of the inner ring of Jordan's advisers, as the minister strove to maintain Britain's ascendancy in China during World War I.[45]

However, though rooted in the nationalisms of the treaty system, the gabelle was far from being an agent of European imperialism, and its members had the outlook of international civil servants. Dane unhesitatingly championed departmental interests against legation policy, if the two happened to conflict.

For example, in April 1916, at a critical point in Yüan Shih-k'ai's fight against the southwestern rebellion, the Japanese government refused to allow its bank to follow Dane's instructions to make releases of salt revenue to Yüan's government, and, within a month, the French, Russian, and British governments followed suit on the grounds that Japan could not be opposed in wartime. Dane protested strongly against what he regarded as both a breach of the loan agreement and a blow against the nonpolitical character of the gabelle: "the withholding of money, which ought to be released and made available for the general purposes of the Chinese Government in accordance with the provisions of the Reorganisation Loan Agreement, merely because it may be thought to be politically expedient to withhold it, appears to me to be wholly unjustifiable if it is not actually dishonourable." The attitude of the powers undermined his position of impartiality: "I feel that I have a strong personal responsibility for the manner in which the Reorganisation Loan has been understood and acted upon by the Chinese Government," [46] and his immediate resignation was averted only by a letter from Sir Edward Grey. Despite his protests, however, the funds were not released; ironically, Dane's only ally in this controversy was Germany, with whom his own country was then at war.

This was not an isolated episode. When Dane visited the Foreign Office in April 1919 on his retirement, Max Muller, head of the Far Eastern department, commented to Jordan: "I wish he were a little more loyal to the Legation and H.M. Government. He is indeed more Chinese than British in his views," [47] though in fact his position could better be described as politically neutral. Dane's last contact with China illustrated this political neutrality. In 1922, a dispute arose between Britain and China over who should succeed Gamble as head of the gabelle, China backing the claims of F. C. Rose of the Chihli River conservancy against those of W. R. Strickland, chief secretary of the salt administration and supported by the legation. According to newspaper reports, Dane, "for whose opinion the Chinese Government entertains the highest respect," was appealed to in his retirement in Herefordshire and, thanks to his reputation for disinterested concern for the salt administration as such, his candidate, Sir Ernest Wilton, former consul-general at Hankow, was eventually accepted by both parties.[48]

Essentially, Dane, like Hart before him, was a departmentalist. Political considerations were always secondary to him, though in general his policy worked in favor of the central government as opposed to regional autonomy. For example, if he had succeeded in releasing $5,500,000 of salt revenue to the Chinese government in April 1916, it might have altered the balance of power in Yüan's favor. Dane remarked to Jordan in June when the money was finally released: "If you had done this three months ago the old man would have been here still." [49] The needs of the organization always came first: to preserve its integrity, he was willing to work with the warlords, just as Sir Frederick Maze later tried to continue the Maritime Customs under Japanese occupation.[50] His priorities were those of a civil servant, whose prime responsibility was for the smooth working of his own department, and this outlook was shared by the inspectorate as a whole.

Administrative branch. Under the system set up by the reorganization loan agreement, Dane's power over the administrative branch was limited. His control of the revenue enabled him to place its salary structure and expense accounts on a properly audited basis; but he had no control over appointments, and, in regions where merchant monopoly still existed, he could not altogether prevent the power to grant such a monopoly from being a source of private profit to the official or to those who appointed him. Nevertheless, in spite of the survival of these traditional practices, the two halves of the gabelle were increasingly assimilated in personnel and orientation.

In organization, the administrative branch was itself becoming more of a specialist agency. Officials were now moved from one salt job to another, suggesting that the salt posts were ceasing to be part of an undifferentiated *cursus.* For example, Yao Yü, who succeeded Chang Hu as Liang-huai commissioner in 1913, was in May 1915 appointed to Kwangtung; again, in 1920, Chin Ting-hsun, salt commissioner in Fengtien, was transferred to the same post in Ho-tung.[51]

In the area of personnel too, the interchange of staff between inspectorate and administration helped to bring about assimilation of the two services. In Szechwan in 1919, for example, Chang Yin-hua, former Ch'uan-nan district inspector, was appointed salt commissioner, the report noting that "Mr. Chang's previous experience in the District Inspectorate facilitated the introduction of various reforms during the year." In 1921, the Ch'uan-pei inspector, C. H. Shui, was appointed to act concurrently as deputy salt commissioner, "the conduct of business between the two departments being greatly facilitated, and numerous important reforms instituted." [52] In the other direction, at Ichang in 1922, the experiment was made of the transportation officer, an administrative official, acting also as assistant inspector, and in Chin-pei a former transportation officer was appointed chief inspector's deputy. These interchanges of personnel were only straws in the wind, but they pointed in the direction of the unification of the two services which was effected by the organic law of 1936.

In political orientation also, the two services were increasingly coordinated: the depoliticization of the inspectorate entailed a parallel divorce from politics for the administrative branch. In Chapter 1, the situation at Ichang was used to illustrate the distortion which involvement in politics imposed on the institutions of the gabelle: for at Ichang in 1900, four different agencies collected tax on salt passing from Szechwan to Hupei (see Table 8). By 1920, this situation had been completely rationalized. There was now only one tax-collecting agency, the transportation office (*chüeh-yün chü*) under the supervision of an assistant district inspector. All tax collected by the transportation officer was remitted via the Bank of China to the group banks at Hankow and thence to Shanghai, and no interference by local authorities was permitted.

The strength of the new system was shown in 1922 when Pi Wei-yuan, transportation officer at the time, tried to subvert it for political ends, with the support of the high commissioner of the Liang-hu, Wu P'ei-fu.[53] Pi endeavored to secure that all revenue, including an illegal surtax of $1.00 a picul known as "foot strength fee," be paid into a private bank for transfer to Wu's headquarters at Wuchang and that a monopoly of the import of salt from Szechwan be established under military control. Although Wu was at that time the strongest warlord in China, the central salt administration was able to frustrate the design: the Bank of China resumed its functions, "foot strength fee" was abolished, the attempted monopoly was abandoned, Pi was dismissed, and his bank became insolvent. The Ichang collectorate was no longer to be enmeshed in the politics of the middle Yangtze.

Conclusion

This account of the depoliticization of the gabelle completes the analysis of Dane's reform. The balance sheet of change, although containing negative items, is, as a whole, positive: a real, if limited modernization of the salt administration had been effected between 1900 and 1920 through the joint efforts of Dane and the Chinese reformers, acting in the favorable circumstances of railway development and increasing centralization. Jordan complacently commented: "The Chinese may curse us for opium, but if they have any gratitude, they will long bless us for salt."[54]

Yet if the only significance of Dane's reforms was as an example of successful synarchy, the study of it would be no more than treaty port antiquarianism, and of limited historical interest. For Chinese history, the significance of the reform lies in its contribution to the peculiar pattern of the modernization process.

On the one hand, salt reform, like other developments in social and administrative institutions, was to provide precedents and personnel for further reform. Encouraged by Dane's success, Thomas W. Lamont's consortium after World War I sought to secure a new massive loan to China on a reorganized tobacco and spirits excise administration. The example of the gabelle influenced Kuomintang institutions also. According to Tseng Yang-feng who worked with both men, T. V. Soong's financial reforms after 1928 may be accurately described as the generalization of Dane's aims and methods, since it was Dane who first demonstrated China's ability to develop adequate internal sources of revenue without deficit or borrowing. The ministry of finance in Soong's time drew many senior personnel from the salt inspectorate; so did the Chinese side of the Kemmerer commission. Even Dane's method of avoiding fraud in the payment of the salt police was later copied by the Chinese army.[55] An example of modernization had been set.

On the other hand, the modernization of the salt administration contributed to a less healthy situation: an increasing imbalance between political and social achievements in republican China. Politically, the record of the early republic was one of failure: the successive failures of Yüan Shih-k'ai, Tuan Ch'i-jui, Wu P'ei-fu, and Chang Tso-lin. Social and administrative institutions per contra were making considerable headway, though so long as political chaos continued, the modernization process was bound to be incomplete in individual institutions and fragmented as regards the country as a whole. Wu Lien-teh, for example, had to wait until 1930 for the Manchurian plague-prevention service to be generalized into a national quarantine service which would deal with all epidemic diseases on a nationwide scale. The salt administration had to wait until the 1950's for a government which would promote the development of a chemical industry, complete the railway network, and eliminate the last vestiges of bureaucratic corruption.

This disequilibrium between social and political development, it may be suggested, was one of the dynamics of Chinese history between 1900 and 1949: it disenchanted men with existing politics and created a reservoir of educated discontent on which revolutionary political movements could draw. The tension between what was institutionally not only desirable but now practicable, and what was actually achieved by the government, could only bring politics into disrepute. If Jordan could argue that "China's present helpless condition is due to the corruption and incompetence of her ruling classes and the exploitation of her weakness by Foreign Powers," [56] many Chinese would draw the same conclusion. Ching Pen-po was contemptuous of the politicians of his generation, while Ma Yin-ch'u, the salt reformer of 1928, although a member of the Legislative Yüan, became an opponent of the Kuomintang and remained on the mainland after 1949.[57] By 1920, the expectations and environment of many urban Chinese were modern; the nation was not. It was not the complete absence of modernization in China which created the potential for revolution, but the modernity of its social institutions as contrasted with the ineptitude of its politics. Of these modern institutions, Dane's salt administration was one of the most successful.

8 | Synarchy

Beyond its Chinese significance, Dane's reform is relevant to broader fields of study: to the character of European imperialism in Asia, and to its role in the modernization of the traditional Asian empires. First, the record of Dane's reform highlights the importance of synarchy, the minimal form of imperialism consisting of European participation in the administration of otherwise independent states. This administrative lend-lease was the predominant instrument of European imperialism in Asia from the sixteenth century to the 1920's for reasons grounded in the intellectual foundations of modern Europe. Second, to narrow the enquiry to fiscal institutions, it may be argued that synarchy of this kind made a dynamic and on balance positive contribution to modernization by sweeping away the financial obstacles to further development.

Synarchy in Asia

Of the various forms of European contact with Asia, synarchy was the most limited and self-restricting. The integrity and independence of the Asian state were left untouched; indeed, a frequent object of synarchy was to reinforce them. What was involved was European participation in certain functions: military, financial, educational, or scientific. This participation might affect the activities of government, but the political process as such was left undisturbed.

The earliest form of synarchy was hardly an imperialism at

all. In the sixteenth century, there was a continuous outflow of talented men from Christendom to the Islamic countries of the Mediterranean. A significant part of the Turkish imperial bureaucracy, for example, was recruited from such renegades, refugees from the aristocratic society of Christian Europe attracted by the greater rewards and mobility of the Ottoman opportunity state. Thus Sokollu Mehmet Paşa, grand vizir from 1565 to 1579, came from the region of Ragusa; Cigala, commander of the Turkish fleet at the end of the century, was a Sicilian; Euldj Ali, *beylerbey* of Algiers and later *kapudan-paşa,* was a Neapolitan; and his predecessor Hassan was a Corsican.[1] These men brought with them the advanced military and administrative technology of Western Europe. The sixteenth-century traveler Philippe Fresne-Canaye wrote in 1573: "Les Turcs ont, par les renégats, acquis toutes les supériorités chrétiennes." [2] Braudel has gone so far as to suggest that the drying-up of this flow of renegades in the seventeenth century was a major factor in the decadence of the Ottoman empire.[3]

It might be questioned, however, whether renegation was truly a form of synarchy. The first clear example of European participation in an oriental administration was the Napoleonic regime in Egypt, 1798–1801.[4] The type of reform instituted in the Ottoman financial system by Jean-Baptiste Poussielgue, administrator general of the finances of Egypt, foreshadowed the more developed synarchy of Dane's day. A dyarchy of French and Egyptian financial officials was established and tax farming was replaced by direct collection paid on a commission basis, reforms which created the fiscal infrastructure for the modernizing regime of Mehemet Ali after the French. Napoleon also sent military advisers to Turkey and Persia in the period before Tilsit in preparation for a combined campaign against Russia. With the agreement with the tsar, these initiatives were discontinued, but French officers played a significant role in stiffening Turkish resistance against the British naval offensive in the Dardanelles in 1807.[5]

If synarchy as a diplomatic instrument was primarily a French invention, its subsequent elaboration in the nineteenth century was mainly British. In the period following the Crimean war, Britain sought to make the Ottoman empire more defensible through internal reforms. Sir Henry Bulwer, British ambassador to Constantinople, 1858–1865, developed a systematic philosophy of synarchy. Turkey, he argued, must be modernized by Europeans who would train Turks, particularly in the fields of finance, army, navy, telegraphs, and public works: "Let the Sultan do as Peter the Great did, as even Mehemet Ali did: let him invite Europeans on honourable terms into his service; let him make use of them in honourable and responsible situations, and in a little time Turkey would be sufficiently Europeanized to look Europe fairly in the face." [6] Bulwer accomplished little of what he projected, but the views he expressed, particularly as regards finance, underlay subsequent synarchies both in the Levant and beyond it: "Without financial order there is no possibility of administrative improvement; with financial order it is impossible for a State to be wholly misgoverned." [7] Reform in Asia was thought to lie through fiscal rationalization.

The fullest development of synarchy came with the high tide of European imperialism, 1880–1925. Of the four Asian empires — the Ottoman empire, Persia, India, and China — only India became a European colony in the sense of being administered, directly and indirectly, by European officials. The other three were set on the road to modernization by synarchic institutions, among which financial agencies played a central role, overshadowing in importance the various military, educational, and technical missions operating intermittently throughout the period. In the Ottoman empire, the administration of the Ottoman public debt, controlling one-third of the revenues of the empire, was set up in 1881; while in 1908, Sir Richard Crawford, formerly of the Egyptian customs, took charge of reforming the Turkish customs service. In Persia, the whole financial administration came under the control of two American financial missions: that of W. Morgan Shuster, treasurer-general of Persia, in 1911, followed by that of Arthur C. Millspaugh, administrator-general of the finances of Persia, 1922–1927. In China, the salt gabelle and the Maritime Customs each collected nearly half of the revenue of the new republic. In the first three decades of the twentieth century, financial synarchy was the principal vehicle of European imperialism on the Asian mainland.

These institutions, operating in diverse circumstances, might in their practice have had little in common except the fact of European participation. In fact, their published reports reveal a high degree of structural similarity in that all of them aimed at Dane's four basic reforms: the creation of a new, fully salaried, and non-political administrative cadre; the centralization of revenues; the extension and intensification of administrative control; and the expansion of trade through its liberalization.

In Persia, for example, Millspaugh argued that "A well-regulated, non-political and impartial finance administration is the best advertisement for Persia abroad." [8] His predecessor Shuster stated that "My purpose in drafting the law of June 13, 1911 was to establish a central organization to be known as the office of the Treasurer-general of Persia, which should be responsible for . . . all revenues and Government receipts, from whatever source derived." [9] He claimed that under his leadership, "the Treasury did succeed in collecting, during a period of five months . . . more of the internal taxes, or *maliat,* than the Government had, so far as the records showed, been able to collect during the entire year preceding our arrival." [10] Millspaugh's struggle with the local opium trade around Isfahan recalls Dane's campaign to tax short-radius salt distribution around Shanghai, while his reference in his ninth quarterly report to "the close control now being exercised over the movement of the opium sap" [11] is again reminiscent of the success stories of the inspectorate reports in China. Like Dane, Millspaugh was a free trader: "A monopoly is bad in principle and usually unnecessary in practice." [12]

A closer parallel to Dane's work may be found in the Ottoman public debt administration during the decade before World War I. In Turkey, as in China, the principal revenue directly administered by foreign officials was a salt tax. The reports of Sir Adam Block, British representative on the council of the administration, 1903–1929, reveal problems and achievements similar to Dane's: obstructive monopolies, doubling of central government revenue, expansion of trade through modern transportation. *Mutatis mutandis,* Block's words in 1910 might have been Dane's in 1918: "Thanks to the control of our Inspectors, the more or less automatic extension of the dépôt service, the development of the railway system and certain police measures taken along the coast of Trebizond and elsewhere, to prevent smuggling, the sale of Salt in the Salt works and dépôts has increased from 150,000 tons in 1882/3 to 230,000 tons in 1906/7." [13] Block, too, expressed warm gratitude for the enlightened support of Djavid Bey, Turkish minister of finance, just as Dane was to express his appreciation of Chang Hu's collaboration.[14]

Block, Shuster, Millspaugh, and Dane therefore formed a group pursuing common ends by similar means, all of them acting on the belief expressed by Shuster that "Western man and Western ideals *can* hustle the East, provided the Orientals realize that they are being carried along lines reasonably beneficial to themselves." [15]

That synarchy rather than colonialism was the predominant form of European imperialism in Asia was due to a variety of factors: rivalry between European powers for control in particular areas and inability to agree on partition; reluctance to assume the commitments, financial and strategic, of colonial administration; the need to find a compromise between advocates of forward policy and critics of imperialism. In addition, while the Asian empires were still militarily vigorous and made good use of their diplomatic opportunities, the partly nonnational character of their government made it easier to insert foreigners into their bureaucracies.

The most enduring reason for the predominance of synarchy, however, lay in the intellectual presuppositions which accompanied the European expansion. On the one hand, since European imperialism originated in a complex of nations partially integrated by their universalistic culture rather than in an imperial state like China where culture and power were identical, the cultural transformation of Asia could be kept conceptually distinct from its political subjection. Indeed, as the Jesuits in China quickly saw, too close an association with European political power might prove a hindrance to their cultural mission.

On the other hand, the old scholastic concept of the ubiquity of natural talent encouraged the more enlightened of the European travelers to admire what they found in the institutions of Asia and to resist the imposition of wholesale change for its own sake. Here again the Jesuits played a central role, the accommodation policy which they adopted being from the start opposed to any drastic revision of Asian culture and institutions.[16] Reconstructed by the Enlightenment and kept alive, as already shown, in Napoleon's Islamic policy, this more generous outlook was never entirely lost. Synarchy, it may be argued, was in effect a secular version of the accommodation policy. Thus Dane genuinely admired the sophisticated institutions of Huai-nan and Szechwan and could be accepted into an indigenous tradition of Chinese salt reform,[17] while Block and Shuster believed that Turkey and Persia could be modernized on their existing foundations. Not only was conquest not required, but on the cultural-institutional level change might best take place in continuity with the past.

In the last resort, therefore, the choice of synarchy as the preferred instrument of European imperialism in Asia is explained in terms of the European distinction between political authority and cultural influence and the willingness of Europeans to accord a positive value to Asian institutions.

Synarchy and Modernization in Asia

In a memorandum written on the eve of the Washington conference of 1921–1922, F. T. A. Ashton-Gwatkin, a senior policymaker at the Foreign Office in London, wrote: "The effect of foreign influence in the salt administration has been wholly beneficial," and again, "The present system appears on the whole to have been satisfactory in its working, and to have procured substantial benefit for both Chinese and foreign interests." [18] In Turkey, Sir Telford Waugh, former consul-general at Constantinople, wrote in 1930 of "the influence of Sir Richard Crawford, who in the six years of his work in Turkey had made himself a name which is still gratefully remembered." [19] To imperial Europe, no doubt, synarchy was satisfactory enough, but to justify these optimistic evaluations from an Asian point of view requires a further assessment, taking more account of local situations, in particular of the political circumstances of the administrations under discussion.

It might be urged that whatever the achievements of bodies like the salt gabelle, the Persian financial mission, and the Ottoman public debt administration, the benefits of synarchy were inevitably outweighed by its disadvantages. In the first place, synarchy might be held to inhibit indigenous reform movements whose programs, being of local origin, might be better suited to the circumstances than formulas imparted ready-made from Europe. Second, even if synarchy was the best way of reforming particular institutions, the European presence might have had a blighting effect on native reformers working in other institutions: it is arguable that no community can modernize itself without full sovereignty and evident control of its own affairs. Finally, even if synarchy did no immediate harm, was it not inevitably, whatever the intentions of its founders, a first step towards full colonial government, direct or indirect? Colonialism in turn would have to be reversed for full modernization to be established.

Dane's reform of the Chinese salt administration is a test of these arguments, either confirming or undermining their validity in at least one concrete instance.

The reorganization loan agreement and the establishment of the foreign inspectorate certainly made it impossible for Ching Pen-po and his friends to come to power in the salt administration. In addition, there are grounds for believing that the *étatiste* approach of the Chinese reformers was more in accordance with China's long-term needs than Dane's policy of minimum state intervention. The experience of other countries in the field of salt administration shows that Dane's path to modernization was not the only one. Japan successfully operated a system of official acquisition not dissimilar to Ching Pen-po's. Austria, where the problems of the Wieliczka mine in the interior of Europe paralleled those of the Szechwan wells in the interior of China, had a system of complete government monopoly. In Brazil, which faced difficulties of long-distance transport by river and rail from the coast to the interior similar to those of China, the Instituto Nacional do Sal, established by the law of June 10, 1940, intervened actively in the trade by fixing production quotas, profit rates, and prices.[20]

Nevertheless, it may be suggested that in the circumstances of early republican China, the practical alternative to foreign reform was not Chinese reform of the salt administration, but no reform at all. Chang Chien and Ching Pen-po were not likely to win their battle with Chou Hsüeh-hsi for Yüan Shih-k'ai's support. Even if they did win and obtain the opportunity to remodel the gabelle as they wished, their scheme was complicated and expensive and might well have collapsed through corruption and the opposition of the same vested interests which frustrated Dane's free trade plans. Tseng Yang-feng believes that though Ching Pen-po had some sound ideas for reform, he was on the whole somewhat unrealistic in approach and lacked Dane's sure grasp of the possible.[21] The experience of H. H. Kung and T. C. Chu, who introduced an ambitious system of government purchase and transportation in 1942, indicates that such schemes could easily remain facades behind which the existing trading patterns continued.

The second criticism of synarchy is more fundamental. It is that the Western presence in institutions such as the gabelle, the Maritime Customs, and the post office, whatever its beneficial effect, aborted the thrust of modernization generally in the period following the revolution of 1911. Within the treaty system, it is held, modernization was impossible, for national sovereignty is its *sine qua non.*

Two answers may be made to this criticism. In the first place, it is difficult to point to anything concrete outside the salt administration which Chinese reformers were prevented from doing by the presence of the forty-one foreigners within it. The tasks generally regarded as essential for China at this time — establishment of central government authority, demobilization of surplus military, completion of the Canton–Hankow railway — were either aided by the success of the foreign gabelle or lay outside its range of repercussion. Moreover, even if the foreign presence had a psychologically discouraging effect on Chinese endeavor in some directions, this was counterbalanced, as Hou Chi-ming has shown, by its retaliation, imitation, and linkage effects in others.[22] The salt industry itself provides an instance of this with the development of the Chiu-ta enterprise at Tientsin to manufacture foreign-style refined salt and its subsequent branching out into soda production in competition with Brunner Mond.

Second, to regard national sovereignty as essential for modernization is to underestimate the possibilities of piecemeal reform, institution by institution, but even if modernization cannot be completed by inches and drops and requires some total change as well, it is not clear that this would lie in the field of constitutional law and the juridical definition of national sovereignty. It is more likely to lie in the field of axiology, in some *Umwertung aller*

Werte, which may or may not accompany resumption of national sovereignty. This criticism of the treaty system looks like an alibi for some other failure.

The final disadvantage of synarchy, it is claimed, was its inherent tendency to develop into full colonialism, an outcome inimical to modernization. Without arguing the apodosis, it may be agreed that in Egypt, Tunis, and Morocco there was a progression from synarchy to indirect colonial rule. Further, in China, Dane's success certainly encouraged the initiation of other projects of synarchy. In the period following World War I, Jordan advocated the unification of China's railways as a basis for national unity, and it was understood in London that the Chinese "had gone so far as to intimate their readiness to have the administration controlled by an organisation similar to the Salt Administration, *i.e.,* an organisation with an influential foreign element." [23] Similarly, a Foreign Office memorandum on currency reform in 1921 concluded: "It is very desirable to induce China to accept the foreign assistance, without which no permanent reform seems possible." [24]

Nonetheless, these projects stopped short of full colonialism and were carefully distinguished from it by the diplomats. Any suggestion of a takeover on the Egyptian model was emphatically rejected. In a memorandum on the eve of the Washington conference, Sir Victor Wellesley, a senior Foreign Office official, argued: "Intervention is not a new idea, but it has been consistently and rightly condemned as both impracticable and impolitic." On the other hand, synarchy in China was different: "If she can be induced to solicit foreign control over certain branches of her Administration, so much the better, as in fact she did in the case of the Salt Gabelle." [25]

The foreign salt gabelle therefore did not prevent a more appropriate reform under Chinese leadership; nor did it obstruct modernization in other fields; and it did not of itself evolve into a colonial administration. In this one concrete instance, the potential disadvantages of synarchy failed to materialize.

The advantages of synarchy, on the other hand, are best seen if the discussion is limited to fiscal institutions, not only in China but also in Persia and in the Ottoman empire, over an extended period of time. The reforms these financial administrations introduced were of lasting significance to the whole structure of the states concerned, conferring the twin benefits of consolidated revenues and a modern civil service, without which no modernization could have taken place.

Bureaucratic monarchy, whether under the Ch'ing, the Ottomans, or the Habsburgs, was a political organism of a higher order than other contemporary forms of government. Rationality in personnel selection, regularity and uniformity of operation, size of territory, concentration of ultimate authority at the center — all these were achieved in a new degree. Still unsolved, however, was the problem of finance. The bureaucratic monarchies of both Europe and Asia suffered from chronic fiscal anaemia, and their histories were punctuated by bankruptcies, devaluations, forced loans, confiscations, and the sale of rank and office. The reason was an apparently vicious circle. Since the state could not afford to pay its officers an adequate salary, it permitted them to remunerate themselves through fees, commissions, perquisites, and irregular exactions, levied with its knowledge but beyond its control. This system produced a discrepancy between gross tax collections from the public and net government receipts, so that whatever the rates of tax, the state did not receive a revenue sufficient to pay its officials.

Sir Henry Bulwer described this syndrome in a letter to Lord John Russell on November 12, 1860: "Your Lordship, however, is perfectly right in considering that the main and principal evil of the Turkish Administration . . . is the corruption which generally pervades it, and which renders the taxation at once grinding to the people and insufficient for the exigencies of the state." However, "Alter this state of things . . . and you alter naturally the basis of the present Budget, and the State will have to charge itself with a far greater expenditure as the only means by which it can possibly, though not certainly, procure a larger revenue." [26]

The remedy for this cycle of insufficient revenue producing a corrupt bureaucracy producing an insufficient revenue was an incorrupt, salaried civil service which would abolish the discrepancy between gross collections and net receipts and thus increase central government revenue sufficiently for the state to be able to afford the new bureaucracy. Higher revenues created states of greater power and versatility. Resources could be channeled into institutions to promote economic development and thus taxable capacity, so that an accelerating cycle of more revenue, greater wealth, more revenue replaced the traditional vicious circle.

Strangely, this crucial breakthrough was not made in Asia where premodern bureaucracy reached its highest form, but in Europe where bureaucratic monarchy was late, fragile, and incomplete. The first tentative moves towards the concept of a salaried civil service — Paul IV's reform of the datary, the anonymous reform project in the Hargrave manuscripts, the Cromwellian attempt "to substitute salaries for fees in the excise, customs, and navy" [27] — all emerged from the court and country conflicts of early modern Europe: the Reformation crisis and the general crisis of the seventeenth century.[28] It may be supposed that it was precisely the vulnerability of the European bureaucracies and the strength of the pressures against them that forced the evolution of a new administrative system which, although applied hesitatingly and reluctantly at first, eventually transformed the character of Western government.

In Asia, on the other hand, the bureaucratic courts were too deeply entrenched for there to be any alternative to them. There was no court-country conflict and no new administrative system; only anarchic outbursts and dynastic successions. The early Ch'ing might consciously avoid the mistakes of the extravagant, fastuous court of the late Ming. The Yung-cheng emperor might limit official corruption with his system of additional stipends. But reform stopped short of its logical conclusion of a salaried civil service. Ch'ien-lung preferred to control corruption by participating in it, channeling it upwards through the patronage empire of his "cormorant minister" Ho-shen. Similarly, the Köprülüs confronted the financial problems of the Ottoman empire in Turkey not by administrative reform but by terror and confiscation, and in Egypt by accelerating the turnover in tax farming by playing off one Mamluk clique against another, while in Persia, Nadir Shah looked to war to restore his finances.

None of the Asian monarchies spontaneously produced the new kind of administrative system,[29] and their finances became ever more inadequate as they had to face not only their own traditional costs but also the cost of modernization. Shuster made this point as regards Persia: "It might have been possible in the past to create a strong central government, without sound financial operations —

indeed, several of the old Shahs succeeded in maintaining a strong control throughout the Empire — but in recent years the time had gone by when Persia could be put in order except through an efficient handling of her taxation and other financial problems." [30]

Thus was created the opportunity for synarchy. What the Asian monarchies could not do for themselves, European financial advisers could do for them. In the letter to Russell on corruption quoted above, Bulwer argued: "I am convinced that if it is to be dealt with effectually, Europeans must be employed in the Financial Administration of Turkey, and given complete control both as to the persons they should employ, the salaries they should give, and the punishments they should inflict." Not only would an enlarged revenue result, but "a new class of functionaries might be formed; whilst the honesty and regularity thus established in one branch of the Administration would introduce, eventually, a change in the others." [31]

Fiscal synarchy, therefore, offered the Asian monarchies the financial resources with which to consolidate their authority and undertake modernization. By and large, advantage was taken of this opportunity. The governments of the Committee of Union and Progress, of Reza Shah Pahlevi, and of Yüan Shih-k'ai, all of whom employed foreign financial advisers, were stronger and more genuinely in command of their countries than their immediate predecessors.

In addition, while it is not always possible to specify how particular revenues were spent, a number of modernization projects were charged directly on funds raised under foreign supervision. Donald C. Blaisdell, the historian of European financial control in the Ottoman empire, notes that "Under the aegis of the Public Debt Administration railroad construction in Turkey received its greatest impetus." [32] Routes financed by it included the Salonika–Constantinople, Salonika–Monastir, and Eskişehir–Konia lines. In Persia, Millspaugh established the sugar and tea monopoly to finance "a trust fund for the railroad construction," [33] so that the Trans-Iranian railway, one of Reza Shah's greatest achievements, was built without foreign capital. In China, the reformed salt administration was a principal contributor of funds to the Chihli River commission, another Sino-foreign body set up in 1918 to undertake a radical attack on the twin problems of the Yungting inundations and the silting of the Hai-ho. Jordan told Curzon: "The work which the Chihli River Commission is doing is one of the greatest importance to the well-being and prosperity of one of the foremost provinces and treaty ports of China." [34] The fruits of foreign financial administration were many and varied.

Conclusion

In the long run, nationalism, whether traditional or modern, could not accept synarchy. The Ottoman public debt administration was phased out by Ataturk, Reza Shah Pahlevi dispensed with Millspaugh once he was firmly in the saddle, and while the foreign salt gabelle continued under the Kuomintang with American leadership replacing the British, the foreign chief inspector's authority was much reduced from what it had been in Dane's day. O. C. Lockhart, the foreign head, resigned in 1941, and the other foreign members of the inspectorate followed during the Pacific war.[35]

Yet once disengaged from its original imperialist milieu, synarchy may not be without relevance for the future. The ongoing modernization process produces uneven advance as between countries and hence a continuing need to rectify this disequilibrium. Specialist agencies, like those of the United Nations or the European community, transnational in organization yet accepting the ultimate authority of local sovereignty, diffusing expertise yet avoiding politics, might provide part of the answer. If so, such institutions should admit a degree of kinship with synarchy and men like Sir Richard Dane.

Appendix

Notes

Bibliography

Glossary

Index

Appendix

Chinese Equivalents for English Terms Used in the Text

Artificially evaporated salt	chien-yen
Cargo certificate	ts'ang-tan
Controller-general of the salt gabelle	tu-pan yen-cheng ta-ch'en
Delivery permit	fa-yen chao
Department of salt administration	yen-cheng yüan
Free trade	min-yün min-hsiao, tzu-yu mao-i
Huai-nan general office	Huai-nan tsung-chü
Local consumption area	shih-an
Merchant transport	shang-yün
Monopoly merchants	yin-shang, p'iao-shang
Official acquisition	kuan-shou
Official supervision and merchant operation	kuan-tu shang-pan
Official supervision and merchant sales	kuan-tu shang-hsiao
Official transport	kuan-yün
Official transport and merchant sales	kuan-yün shang-hsiao
Official transport and official sales	kuan-yün kuan-hsiao
Official transport office	kuan-yün chü
Owls, owling	hsiao
Passport	hu-p'iao
Provincial distributor	p'u-fan, shui-fan, Hu-shang, an-shang
Provincial sales office	tu-hsiao tsung-chü
Sales suboffice	tu-hsiao fen-chü
Salt administration	yen-cheng, yen-wu, yen-wu shu
Salt commissioner	yen-yün shih, yen-fa tao
Salt controller	yen-yün shih
Salt division	yin-ti
Salt receiver	ch'ang-kuan

Salt taotai	yen-fa tao
Solar-evaporated salt	shai-yen
Szechwan likin office	Ch'uan-yen tsung-chü
Taxation at source	chiu-ch'ang shou-shui, chiu-ch'ang cheng-shui
Taxation at source and minimal interference	chiu-ch'ang cheng-shui jen ch'i so chih
Transport merchants	yün-shang
Transport permit	ch'ung-yen chih-chao
Wastage allowance	lu-hao
Yard	ch'ang
Yard merchant	ch'ang-shang
Yin system	yin-chih

Notes

Abbreviations Used in the Notes

CSL	*Ta Ch'ing li-ch'ao shih-lu* (Taiwan, 1963).
Customs Report	China, Imperial Maritime Customs, *Salt: Production and Taxation,* V Office Series, Customs Papers, No. 81 (Shanghai, 1906).
Dane Report	Report by Sir Richard Dane, K.C.I.E., on the Reorganisation of the Salt Revenue Administration in China, 1913–1917.
Decennial Reports	*1902–1911.* China, Imperial Maritime Customs, *Decennial Reports 1902–1911,* 3 vols. (Shanghai, 1912).
Decennial Reports	*1912–1921.* China, Maritime Customs, *Decennial Reports 1912–1921,* 2 vols. (Shanghai, 1922).
District Reports	Reports by the District Inspectors, Auditors and Collectors on the Reorganisation of the Salt Revenue Administration in China, 1913–1917, 1918, 1919–1921, 1922.
NCH	*North China Herald and Supreme Court and Consular Gazette.*
TYTP	*T'an-yen ts'ung-pao* (journal; Shanghai, 1913–1914).
YCTC	*Yen-cheng tsa-chih* (journal; Peking, 1912–1915).
YWKMS	Ching Pen-po, *Yen-wu ko-ming shih* (Nanking, 1929).

Reign Periods

HT	Hsüan-t'ung period, 1909–1912.
KH	Kuang-hsü period, 1875–1908.

1. The Structure of the Salt Administration in 1900

1. Pliny the elder, *Naturalis Historia,* Book XXXI, xli, 88; Fernand Braudel, "Achats et ventes de sel à Venise (1587–1593)," *Annales, economies, sociétés, civilisations,* 16.5:961 (September–October 1961); Herbert Franz

Schurmann, *Economic Structure of the Yüan Dynasty* (Cambridge, Mass., 1956), p. 175.

2. British Foreign Office Records, F.O. 405/201, China, Annual Report 1910, p. 3.

3. *Decennial Reports 1912–1921,* I, 311, II, 53; Wu Lien-teh, *Plague Fighter: The Autobiography of a Modern Chinese Physician* (Cambridge, England, 1959), p. 388; Sidney D. Gamble, *Peking: A Social Survey* (New York, 1921), pp. 125, 75–76.

4. Jordan Papers, Jordan to Langley, November 24, 1914.

5. *NCH,* April 11, 1914, p. 103.

6. *Encyclopaedia Britannica* (London, 1957), "Salt," p. 899; A. and N. L. Clow, *The Chemical Revolution* (London, 1952); A. and N. L. Clow, "The Chemical Industry: Interaction with the Industrial Revolution," in Charles Singer, E. J. Holmyard, A. R. Hall, and Trevor I. Williams, eds., *A History of Technology* (Oxford, 1954–1958), IV, 239.

7. Customs Report, p. 2; F. A. Cleveland, "Statistical Review of the Work of the Inspectorate, 1913–1933, with Special Attention Given to the Evaluation of Results Achieved During the Last Five Years," Chinese National Government, Ministry of Finance, Inspectorate of Salt Revenue, n.d., p. 5.

For salt and fisheries in medieval Europe, see A. R. Bridbury, *England and the Salt Trade in the Later Middle Ages* (Oxford, 1955), especially Introduction and Chapter I; and for salt and the dairy industry, see Edward Hughes, *Studies in Administration and Finance, 1558–1825* (Manchester, 1934), pp. 9, 493–494.

For fishery salt in the Chusan archipelago, see Dane Report, p. 120, and *YWKMS,* p. 3. For Hainan, see Customs Report, pp. 292–295.

8. These are approximate figures only, the evidence being most certain for Szechwan and the central Chinese provinces. In Szechwan salt division, the official quota for the province proper was 3,630,000 piculs, and the actual amount released for sale in 1914 was 4,430,000 piculs (District Reports, 1913–1917, p. 68).

In Hunan, consumption was calculated at 2,496,000 piculs (Customs Report, p. 68) and in Hupei at 2,400,800 piculs (Customs Report, p. 79), the figures being based on the official Huai quotas and on Szechwan salt passing the Ichang salt likin collectorate.

Honan figures are more difficult to establish as the province was divided between four salt divisions: sixty-one districts belonged to Ch'ang-lu, thirty-four to Ho-tung, fourteen to Huai-pei, and nine to Shantung (Dane Report, pp. 75, 87; also Customs Report, p. 155). Gustav Detring, the Tientsin customs commissioner, estimated that 1,500,000 piculs of Ch'ang-lu salt were consumed in Honan (Customs Report, p. 8), and the official Huai-pei quota for the fourteen Honan districts was 198,000 piculs (Customs Report, p. 144). Ho-tung salt division which was estimated (Customs Report, p. 59) to produce 1,650,000 piculs of salt, also supplied fifty-one districts in Shansi and thirty-five Shensi districts besides the thirty-four Honan districts, and so it may well have supplied Honan with 300,000 to 400,000 piculs, making a total for Honan of around 2 million piculs, a likely amount in view of the Hunan and Hupei figures.

In Chihli, Detring believed the consumption of salt (Customs Report, p. 8) to be 3 million piculs, of which 2 million may be estimated to have been consumed in the more thickly populated areas around Peking and Tientsin. Detring's figure is borne out by the earliest inspectorate report (District Reports, 1913–1917, p. 1).

For the Yangtze and West River deltas, the figures are more conjectural since these were areas in which much salt went untaxed. For salt purposes, the Yangtze delta may be defined as the *shih-an* area of Huai-nan (for a definition of this, see Customs Report, p. 157) and the Su Wu-shu area of Liang-che salt division (Dane Report, p. 116). The official quotas for these two were 381,360 piculs (Customs Report, p. 177) and 408,720 piculs (Dane Report, p. 118) respectively, but this was certainly an underestimate for actual consumption. In 1918, when the inspectorate had brought more of the salt consumed under taxation, the figures for the *shih-an* and Su Wu-shu were 1,058,490 piculs and 875,708 piculs respectively (District Reports, 1918, pp. 20, 43). Even this was probably still not all the salt consumed, as the inspectorate never achieved complete efficiency in this area, but it does suggest that the true consumption figure was approximately 2 million piculs.

As for the West River delta, the official quota for Kwangtung proper in 1900 was only 1,099,479 piculs (Customs Report, p. 282), and for the whole Kwangtung salt division, 2,095,149 piculs. Dane, however, regarded Kwangtung as of equivalent standing to Ch'ang-lu (Dane Report, p. 215), and the amount paid in 1911 for the tax farm of the whole monopoly area, 5,800,000 taels (Dane Report, p. 126), supports this view, as the Ch'ang-lu revenue from the official quotas came to approximately the same sum (District Reports, 1913–1917, pp. 1–4). The 1900 official quotas for Kwangtung proper should, then, probably be at least doubled to give a true consumption figure of 2 million piculs.

9. W. W. Rostow, *The Stages of Economic Growth* (Cambridge, England, 1962), p. 52.

10. Customs Report, p. 194; Esson M. Gale and Ch'en Sung-ch'iao, "China's Salt Administration: Excerpts from Native Sources," *Journal of Asiatic Studies,* 2.1:285–286 (June 1959).

11. Alexander Hosie, *Manchuria: Its People, Resources and Recent History* (London, 1901), p. 226; Customs Report, p. 210.

12. Customs Report, p. 209. This report contains the best account of the Huai-nan technique.

13. Chang Chien, *A Plan for the Reform of the National Salt Administration* (Shanghai, 1913), p. 8. For Wolf's translation, salt division, see Customs Report, p. 154; for translation by R. de Luca (customs commissioner in charge of the Hupei salt likin collectorate), salt circuit, see Customs Report, p. 84; and for Dane's translations, administrative division, territorial salt district, and salt administration, see Dane Report, pp. 64, 68, 136.

Yin-ti originally referred probably not to the jurisdiction of officials, but to the legal sales area of the salt, i.e. it was a commercial rather than an administrative term. For example, a memorial of Liang-kiang governor-general Liu K'un-i in 1881 (*CSL:KH,* 130:3) refers to the "Huai sales *yin-ti*" (*Huai hsiao yin-ti*) as if *yin-ti* by itself was not enough and needed further elaboration.

By the twentieth century, however, *yin-ti* was certainly a territorial unit: the Customs Report contains a Chinese map showing the *yin-ti,* and in 1911 Liang-kuang governor-general Chang Ming-ch'i could request that three prefectures in Hunan be transferred to Yüeh salt *yin-ti* (*CSL:HT,* 49:15). Any distinction between commercial and administrative applications had become vestigial.

14. Customs Report, p. 84; Dane Report, p. 90; District Reports, 1919–1921, p. 254.

15. Customs Report, pp. 68–72; *CSL:HT,* 49:15.

16. Ferdinand von Richthofen, *Baron Richthofen's Letters, 1870–1872,* 2nd ed. (Shanghai, 1903), p. 39; Dane Report, p. 17; Customs Report, p. 342.

17. Customs Report, p. 7.

18. Li Jo-ping, ed., *A Collection of Modern Chinese Essays,* "A Visit to Charkan Salt Bridge: Mirages and Sea-markets."

19. Richthofen, p. 59.

20. R. H. Tawney, *Business and Politics under James I* (Cambridge, England, 1958), pp. 25–26; Frédéric Mauro, *Le Portugal et l'Atlantique au XVIIe siècle (1570–1670)* (Paris, 1960), pp. 259–277; Huguette and Pierre Chaunu, *Séville et l'Atlantique (1504–1650)* (Paris, 1955–1959), VIII, Part One, 606–612.

21. For the foreign trade figures, see Stanley F. Wright, *Hart and the Chinese Customs* (Belfast, 1950), p. 892.

The opium valuation has been calculated on the basis of 376,000 piculs of native opium traded (H. B. Morse, *The Trade and Administration of the Chinese Empire* [London, 1908], p. 350) at 340 taels a picul (British Parliamentary Papers, China No. 1 [1908], *Correspondence Respecting the Opium Question in China,* Jordan to Sir Edward Grey, August 14, 1907), while the interregional trade in rice has been estimated at 25,000,000 piculs at 4 taels a picul.

For salt in China's domestic trade, see Hosie, *Manchuria,* p. 239; *CSL:KH,* 474:17; Richthofen, p. 74; Customs Report, pp. 304–305, 343.

22. Michel Mollat, ed., *Le Rôle du sel dans l'histoire* (Paris, 1968), p. 237.

23. Customs Report, pp. 46–48, 160, 186, 204.

24. Customs Report, p. 148; *CSL:KH,* 473:11; *CSL:KH,* 511:5.

25. Customs Report, p. 148.

26. Customs Report, p. 148.

27. Customs Report, pp. 141, 151.

28. Dane Report, p. 241.

29. Dane Report, pp. 150, 80, 35.

30. Customs Report, pp. 182–183, gives a detailed table showing the increment in value of one catty of salt from the time of production to the

time of sale for consumption in Anhwei. The figures for Kiangsi, Hupei, and Hunan would be almost identical. The breakdown was as follows:

Price paid to producer	3.827 cash	= 7.654 percent
Charges incurred in trade	7.299	= 14.598 percent
Taxation	20.124	= 40.248 percent
Profits of merchants	18.750	= 37.500 percent
Retail price	50.000	

Braudel, "Achats et ventes," pp. 964–965, has the following breakdown:

Purchase price	12 percent
Trade expenses	7 percent
Taxation and profit	81 percent

These figures refer to the cities of the Venetian mainland, such as Treviso, Verona, and Brescia. Venice itself was privileged in having low taxation, just as the Yangtze delta region was.

31. Dane Report, pp. 38, 79–80.

32. Customs Report, p. 143.

33. Customs Report, p. 114; Dane Report, p. 82. For the Ch'ang-lu monopolists' foreign debt, see *Decennial Reports 1902–1911,* I, 199–200.

34. Customs Report, p. 204. Bruce, moreover, went on to argue (Customs Report, p. 204): "Smuggling is undoubtedly more prevalent than it ought to be; at the same time I am of the opinion that, like the proverbial 'landlady's cat,' the salt smuggler is made to carry the blame of many more sins than his own." For the illicit trade in Huai-nan, see Customs Report, pp. 117–118, 150; for the Pakhoi-West River route, see Customs Report, pp. 309, 315–319, and Dane Report, p. 216.

35. Customs Report, p. 213.

36. Dane Report, p. 18.

37. Dane Report, pp. 17–18. Dane's "fluvio-dynamic" view of the Chinese salt trade must be distinguished from Wittfogel's hydraulic theory: most of the inland waterways involved were natural and not man-made, and hence did not necessitate the large-scale organization of Wittfogel's theory. They did, however, facilitate control by the bureaucracy. For the effect of this control, see below, Chapter 6.

38. For salt in the Yangtze delta, see Customs Report, pp. 218–219, 107–112, 150, 231, and for the salt quotas in particular, see Customs Report, p. 177, and Dane Report, p. 118.

39. Customs Report, p. 91.

40. Dane Report, p. 227.

41. For multiple heads of assessment, see District Reports, 1913–1917, p. 4; Customs Report, pp. 140, 48–49, 107–109; Dane Report, p. 20. For revenue allocation, see Customs Report, pp. 125, 67.

42. This figure of 24,000,000 taels for gross salt tax collection is well below that of Morse, 64,000,000 taels for 1907 (*The Trade and Administration of*

the Chinese Empire, p. 103), and below those also of most of the customs commissioners in the report of 1901. Alec W. Cross at Yochow argued for 86,-400,000 taels (Customs Report, p. 69); de Luca at Hankow for 43,400,000 (Customs Report, p. 79); Wolf at Ta-t'ung for 36,189,045 (Customs Report, p. 146); and only P. C. Hansson at Nanking argued in favor of as low as 20 million taels (Customs Report, p. 223). The reasons for accepting a low figure for 1900 are as follows:

In 1914, the inspectorate, having consolidated all the old imperial taxes into one, collected the equivalent of 40 million taels tax (Dane Report, p. 276). This was the figure for gross salt revenues given in the imperial budget for the third year of Hsüan-t'ung (*CSL:HT,* 61:4). Salt taxes, however, had been sharply increased since 1900 (see below, Chapter 2), and for the last year of Kuang-hsü their total was given as 29 million taels (*CSL:HT,* 61:3). The inspectorate may not have been 100 percent efficient in 1914, and the imperial budgets certainly left out items which properly should have been regarded as taxes, but in both cases the margin of error would not have been more than 25 percent. A figure of 24 million taels for gross salt revenue before the post-Boxer increases would therefore seem plausible.

A second line of reasoning leads to the same conclusion. In those areas where both sets of figures are available, gross revenue in 1900 runs at between 50 and 100 percent above reported revenue. For example, in Liang-huai, where a revenue of 4,690,000 taels was reported (Customs Report, p. 30), gross revenue was 7,400,000 taels (Customs Report, pp. 67, 117, 127, 142, 167–170, 177–180). At the Ichang collectorate, the two figures were 1,362,000 taels (Customs Report, p. 31) and 2,025,000 taels (see below, Table 8; see also Customs Report, p. 42). In Ch'ang-lu, the gross revenue in 1900 may be calculated at 1,800,000 taels ("Chung-kuo yen-cheng yen-ko shih," *YCTC,* No. 19, *chuan-chien* 2:3–5, 10–11 [August 1915]), against a reported revenue of 1,100,000 taels (Customs Report, p. 29). In Kwangtung province, customs commissioner Paul King stated that of a total revenue of 1,357,000 taels, only 639,250 taels "are reported to the Imperial Government" (Customs Report, p. 258).

If the gross revenue for the whole of China were 50 percent above the reported collection of 13,490,000 taels, it would total 20,235,000 taels; while if it were 100 percent above the reported collection, it would total 26,980,000 taels. 24,000,000 taels, therefore, again seems a probable figure.

It may be noted that Dane himself was not entirely clear as to the distinction between the old reported revenue of 13 million taels and what was genuinely available to Peking (Dane Report, p. 1). For the old regime, however, it is necessary to distinguish the three totals: gross, reported, and available.

43. Customs Report, p. 69.

44. Customs Report, p. 287.

45. Customs Report, p. 339.

46. *CSL:HT,* 61:3–4. The seven Ch'ang-lu salt controllers were as follows: Yang Tsung-lien, *CSL:KH,* 505:5; Wang Jui-kao, *CSL:KH,* 531:12; Lu Chia-

ku, *CSL:KH,* 568:6; Chou Hsüeh-hsi, *CSL:KH,* 584:6; Ling Fu-p'eng, *CSL:KH,* 586:19; Chang Chen-fang, *CSL:HT,* 55:6; Liu Chung-lin, *CSL:HT,* 55:9–10.

47. The *shih-lu* references are as follows: K'o Feng-shih, *CSL:KH,* 473:19; Shih-Chieh, *CSL:KH,* 474:5; Yang Wen-ting, *CSL:KH,* 487:14; Yang Tsung-lien, *CSL:KH,* 505:5; Ch'eng I-lo, *CSL:KH,* 510:5; En-ming, *CSL:KH,* 523:23; Ying Jui, *CSL:KH,* 541:10; Hsin Ch'in, *CSL:KH,* 565:10; Lu Chia-ku, *CSL:KH,* 568:6; Wang Jen-wen, *CSL:KH,* 576:18; Ts'ui Yung-an, *CSL:KH,* 576:22; En-lin, *CSL:KH,* 580:12; Ling Fu-p'eng, *CSL:KH,* 586:19; Lu Hsüeh-liang, *CSL:KH,* 589:7; Hui Shen, *CSL:KH,* 589:16; Chao Pin-yen, *CSL:HT,* 12:10; T'ang Shou-ch'ien, *CSL:HT,* 16:29; Ma Chi-chang, *CSL:HT,* 33:35; Ting Nai-yang, *CSL:HT,* 36:11; Chang Chen-fang, *CSL:HT,* 51:14; Liu Chung-lin, *CSL:HT,* 55:9–10.

48. Dane Report, p. 1; *CSL:HT,* 61:47; Dane Report, p. 147.

49. Customs Report, p. 85. Ch'uan salt in Hupei paid 1,460,000 taels to the Hu-kuang authorities and Huai salt 650,000 taels, before the Anglo-German loan agreement of 1898 complicated matters. After it, the corresponding figures would be 600,000 and 400,000 taels. For Ch'uan revenue, see Table 8 and Customs Report, pp. 42–50; for Huai revenue, see Table 4 and Customs Report, p. 79.

50. *CSL:KH,* 487:11–12.

51. Dane Report, pp. 87–90; District Reports, 1913–1917, pp. 7–11.

52. Attempts to calculate total consumption by multiplying population by some figure for average per-capita consumption (Chang Chien—12 catties a year, de Luca—8 catties, and Wolf—9 catties) are suspect, inasmuch as the Chinese population could reduce its consumption severely if circumstances so compelled: the phrase "the people are eating insipid food" recurs in memorials. In Shantung, however, there was no absolute shortage of salt, as the whole coastline could be made productive, so here the argument may legitimately be employed. Wolf's estimate of consumption per head (Customs Report, p. 145) is safer than Chang Chien's (Chang Chien, p. 5).

53. In Shantung, *yin* and *p'iao* had much the same sense as in Szechwan. *Yin* certificates were used in long-distance waterborne trade; *p'iao* certificates in local, mainly land-borne trade. Some *p'iao* areas however, near the Hsiao-ch'ing River, were supplied by water. See Dane Report, pp. 87–88, and Customs Report, pp. 12–13. Being carried by land was thus not part of the *definition* of *p'iao* salt, only a common *attribute* of it in certain salt divisions.

For the quite different significance of *yin* and *p'iao* in nineteenth-century Huai-pei and Huai-nan, see Thomas A. Metzger, "T'ao Chu's Reform of the Huai-pei Salt Monopoly," *Papers on China,* 16:1–39 (Harvard University, East Asian Research Center, 1962).

In general, *p'iao* certificates covered a smaller quantity of salt than *yin* certificates, hence their suitability as a unit for land transportation, though in Huai-nan the *p'iao* was a larger unit than the *yin* (Customs Report, p. 64; Dane Report, p. 99). See below, Chapter 5, note 19.

2. Change in the Salt Administration, 1900–1911

1. *Decennial Reports 1902–1911,* I, 3, 261, 295–296, 11–15, 244, II, 5.

2. *Decennial Reports 1902–1911,* I, 339.

3. Stanley F. Wright, p. 892; *Decennial Reports 1902–1911,* II, 330–331, 314.

4. *Decennial Reports 1902–1911,* I, 345, 200.

5. *Decennial Reports 1902–1911,* I, 232, 423, II, 212, 311.

6. *CSL:KH,* 529:6; *CSL:HT,* 44:5; *CSL:HT,* 42:22. For the Canton lottery, see also Morse, *The Trade and Administration of the Chinese Empire,* pp. 104–105.

7. In 1900, the only hypothecation of salt revenue was 1,800,000 taels from the salt likin collectorates of Ichang, Hankow, and Ta-t'ung for the Anglo-German loan of 1898.

For the breakdown between the collectorates, see T'ang Hsiang-lung, "Min-kuo i-chien kuan-shui tan-pao chih wai-chai," in Pao Tsun-peng et al., eds., *Chung-kuo chin-tai-shih lun-ts'ung,* Second Series (Taipei, 1958), III, 111.

The salt revenues also served as collateral security for the Boxer indemnity.

8. For the Canton lottery, see note 6 above. For Szechwan, see *CSL:HT,* 2:16.

9. *CSL:HT,* 52:15.

10. "Chung-kuo yen-cheng yen-ko shih," pp. 3–5, 10. The 1900 figure has been arrived at by subtracting the items added after 1900 from the 1911 total.

11. *CSL:HT,* 61:3–4; *CSL:KH,* 483:4.

12. For Honan, see *CSL:KH,* 507:2; for Kweichow, see *CSL:KH,* 509:2; for Hu-kuang, see *CSL:KH,* 592:9; for Liang-kuang, see *CSL:KH,* 512:20; for Liang-kiang, see *CSL:HT,* 23:51.

13. E-tu Zen Sun, *Chinese Railways and British Interests, 1898–1911* (New York, 1954), p. 106. For a list of charges and obligations secured on the salt revenue, see Dane Report, p. 276.

14. *CSL:KH,* 592:9.

15. *CSL:KH,* 561:10.

16. District Reports, 1913–1917, p. 4.

17. *CSL:HT,* 39:29–30.

18. *Decennial Reports 1902–1911,* II, 315.

19. Tientsin *Ta-kung-pao,* March 3, 1955, May 15, 1955, September 2, 1955.

20. District Reports, 1913–1917, p. 102.

21. District Reports, 1913–1917, p. 5.

22. Until the Tao-kuang period, Shantung salt was administered by the Ch'ang-lu authorities. See Gale and Ch'en, p. 284.

23. Dane Report, p. 87; District Reports, 1913–1917, p. 8.

24. District Reports, 1913–1917, p. 12; Dane Report, p. 78.

25. *CSL–HT,* 23:16–17.

26. *CSL:HT,* 10:43. See also *CSL:HT,* 63:31, where the department of salt administration memorialized that Huai-nan merchants were using steamships to import Ch'ang-lu and Fengtien salt.

27. District Reports, 1913–1917, p. 37; Dane Report, p. 101.

28. For an example of the censorate defending local interests, in this case in Szechwan, see *CSL:HT,* 26:18–19, and *CSL:HT,* 40:14.

29. Dane Report, p. 125. The context, a discussion of the Fukien monopoly system set up in 1912, makes it clear that by "a Government monopoly" Dane was in fact referring to "official transport."

30. *CSL:KH,* 481:17.

31. Dane Report, p. 70.

32. For Italy, see *CSL:HT,* 19:10–13; for Japan, see *CSL:HT,* 47:8–11.

33. For Kwangtung, see *CSL:KH,* 529:6; for northern Shansi, see *CSL:HT,* 40:9–11; for Jehol, see *CSL:KH,* 568:11; and for Manchuria, see *CSL:HT,* 12:9–10.

34. *CSL:HT,* 25:16; for Chao's plan, see *CSL:HT,* 24:14–15.

35. *CSL:HT,* 25:15–16; *CSL:HT,* 26:18–19.

36. *CSL:KH,* 572:20; *CSL:HT,* 25:15; *CSL:HT,* 28:18–19; *CSL:HT,* 32:10; *CSL:HT,* 39:35.

37. *CSL:KH,* 572:20.

38. Perhaps on account of the jealousy of other railway promoters such as Sheng Hsüan-huai and Yüan Shih-k'ai. For Ts'en's railway plan, see *CSL:KH,* 572:22, and for his appointment in Szechwan, see *CSL:HT,* 61:34.

39. E. H. Parker, *China: Her History, Diplomacy and Commerce* (London, 1917), p. 243.

40. Letter of Miss D. M. Dane (Sir Richard Dane's niece) of September 11, 1965, to the writer.

41. *CSL:HT,* 26:7–8. See also H. S. Brunnert and V. V. Hagelstrom, *Present-Day Political Organization of China,* tr. A. Beltchenko and E. E. Moran (Shanghai, 1912), p. 122.

42. *CSL:HT,* 32:11; *CSL:HT,* 33:4–5; *CSL:HT,* 26:7–8; *CSL:HT,* 34:20–21; *CSL:HT,* 61:2–7; *CSL:HT,* 61:10.

43. Chang Chien, pp. iv–v. For the budgets, see Chang Chien, p. 39; Dane Report, p. 2; and *CSL:HT,* 61:59–61.

44. *CSL:HT,* 61:2–7.

45. Dane Report, p. 1. Dane implies that this sum of 13 million taels was what was available to Peking. Most likely, however, it was only the reported collection of the Customs Report, and hence in excess of what was genuinely at Peking's disposal. See above, Chapter 1, note 42, and Table 3.

46. *CSL:HT,* 61:51; *CSL:HT,* 62:8; District Reports, 1913–1917, pp. 4, 13, 19–23.

47. In a despatch to Grey in 1912 (British Foreign Office Records, F.O. 371/1321, Jordan to Grey, July 11, 1912, File No. 32388), Jordan stated that Hsiung Hsi-ling, then Chinese minister of finance, had told him that 20,000,000 taels of salt revenue, out of a gross figure of 40,000,000 taels, reached Peking by the end of the Ch'ing period. This 20,000,000 taels was probably

revenue reported to Peking rather than revenue genuinely available to it, since Jordan in his despatch compared it to the old reported revenue of 13,000,000 taels. Negotiations between British banks and Chinese regional authorities for small-scale loans secured on local salt revenues in the months immediately before October 1911 (British Foreign Office Records, F.O. 405/205, Affairs of China, July to December 1911) confirm the fact that the old system of decentralized control was still in force then. An anonymous, but seemingly reliable, author, "La Gabelle du sel," *La Chine,* 40:436 (April 1923), stated in 1923 that until 1911 only 14 million dollars were actually received by Peking: "En 1911, une Administration Centrale du sel a été créée à Pékin qui ne put cependant pas changer le régime existant."

48. Dane Report, p. 3.

49. For Huai-pei, see *CSL:HT,* 39:4–5; for northern Shansi, see *CSL:HT,* 40:9–11; for Yunnan and Szechwan, see *CSL:HT,* 40:11; and for Fengtien, see *CSL:HT,* 41:19.

3. The Salt Administration in the Revolution of 1911

1. *NCH,* April 8, 1911, p. 87.

2. *NCH,* November 4, 1911, p. 303; *NCH,* June 10, 1911, p. 667; *NCH,* June 22, 1912, p. 838.

3. *Decennial Reports 1912–1921,* II, 103; *YWKMS,* pp. 3–4; Dane Report, p. 269: "The smuggling of so-called fishery salt from Shantung, Huaipei and the islands of the Chusan Archipelago to the Provinces on the Yangtze was, and still is, a most lucrative business, and causes a most serious loss of revenue."

4. On the salt problems of eastern Chekiang, see *YWKMS,* pp. 3–6.

5. *NCH,* September 16, 1911, p. 696; *NCH,* October 7, 1911, pp. 37–38; British Parliamentary Papers, China No. 3 (1912), *Further Correspondence Respecting the Affairs of China,* p. 54.

6. *YWKMS,* p. 7; *YWKMS,* Appendix, pp. 10–23.

7. *NCH,* October 21, 1911, p. 164.

8. British Parliamentary Papers, China No. 1 (1912), *Correspondence Respecting the Affairs of China,* p. 8; *NCH,* October 21, 1911, p. 164; Dane Report, p. 146.

9. Dane Report, p. 167; *YWKMS,* pp. 11, 19.

10. *YWKMS,* p. 16.

11. *YWKMS,* Appendix, p. 12.

12. *CSL:KH,* 458:10; *CSL:KH,* 497:18; *CSL:KH,* 584:7–8.

13. *YWKMS,* Appendix, p. 19; *NCH,* October 21, 1911, p. 160; *NCH,* December 9, 1911, p. 663; *Decennial Reports 1902–1911,* I, 393.

14. Dane Report, p. 3.

15. *YWKMS,* Appendix, p. 17.

16. Dane Report, p. 114.

17. Dane Report, p. 102.

18. For the Kiangsu movement, see Samuel C. Chu, *Reformer in Modern*

China: Chang Chien, 1853–1926 (New York, 1965), pp. 114–143. Also see *YWKMS,* pp. 1–9.

19. For the Chekiang movement, see *YWKMS,* pp. 3–6.

20. District Reports, 1913–1917, p. 27; Dane Report, pp. 96, 116.

21. For the Manchurian movement, see Shou P'eng-fei, "Tung-san-sheng yen-cheng kai-ko i-chien-shu," *YCTC,* No. 3, *chuan-chien* 2:1–18 (March, 1913). See also Dane Report, p. 71.

22. *YWKMS,* p. 10.

23. *YWKMS,* p. 7.

24. *YWKMS,* p. 10; Shou, pp. 13–17; *YWKMS,* p. 7; Dane Report, pp. 123–124.

25. *YWKMS,* p. 1.

26. *YWKMS,* Appendix, p. 8.

27. This, at least, was the case in Huai-nan. Customs Report, pp. 66–67: "It is considered a good stroke of business to use a *p'iao* twice in one year; the general average is three times in two years." In Huai-nan, *p'iao* had much the same significance as *yin* in other salt divisions.

28. Customs Report, p. 190.

29. District Reports, 1913–1917, pp. 9, 12; *YWKMS,* Appendix, p. 12; Customs Report, p. 266.

30. *YWKMS,* p. 2; Shou, p. 11.

31. *YWKMS,* Appendix, p. 4.

32. *YWKMS,* Appendix, pp. 2–3; Benjamin Schwartz, *In Search of Wealth and Power: Yen Fu and the West* (Cambridge, Mass., 1964), p. 115.

33. *YWKMS,* Appendix, p. 15; *YWKMS,* p. 2.

34. *YWKMS,* p. 7; Dane Report, pp. 145–146, 126, 123.

35. Dane Report, pp. 76, 157.

36. Dane Report, pp. 3–5.

37. Dane Report, p. 4.

38. Dane Report, pp. 7, 5.

39. P'u Yu-shu, "The Consortium Reorganization Loan to China, 1911–1914: An Episode in Pre-war Diplomacy and International Finance" (Ph.D. dissertation, University of Michigan, 1951).

40. British Foreign Office Records, F.O. 371/1313, Hillier to Charles Addis, February 24, 1912, File No. 8394.

41. British Foreign Office Records, F.O. 371/1317, Hongkong and Shanghai Bank to Foreign Office, April 19, 1912, File No. 16505.

42. P'u, p. 312.

43. British Foreign Office Records, F.O. 371/1591, Jordan to Grey, January 20, 1913, File No. 5468.

44. British Foreign Office Records, F.O. 371/1325, Jordan to Grey, November 26, 1912, File No. 53037.

45. British Foreign Office Records, F.O. 371/1591, Jordan to Grey, February 14, 1913, File No. 7213.

46. British Foreign Office Records, F.O. 371/1591, Jordan to Grey, February 7, 1913, File No. 8761.

47. Jordan Papers, Jordan to Langley, September 17, 1914.

48. For Dane's appointment, see British Foreign Office Records, F.O. 371/1591, 1592, 1593.

49. Haxthausen to Arthur Zimmerman, December 31, 1912, quoted in P'u, p. 362; Krupensky to S. D. Sazonov, November 29, 1912, quoted in P'u, p. 349.

50. P. H. Kent, *The Passing of the Manchus* (London, 1912), p. 370; P'u, p. 515.

51. British Foreign Office Records, F.O. 371/1593, Jordan to Grey, March 31, 1913, File No. 17560.

4. Dane's Reforms, 1913–1918: The Basic Achievement

1. On Chadwick, see S. E. Finer, *The Life and Times of Sir Edwin Chadwick* (London, 1952). On Morant, see *Dictionary of National Biography, 1912–1921* (Oxford, 1927), pp. 386–388. On Stephen, see R. B. Pugh, "The Colonial Office, 1801–1925," in E. A. Benians, James Butler, and C. E. Carrington, eds., *The Cambridge History of the British Empire,* III (Cambridge, 1959), 711–768. On Trevelyan, see especially the bound volumes of Lord Clarendon's letters from Ireland, 1847–1852, in the Clarendon Papers (Bodleian Library, Oxford). Trevelyan was assistant-secretary to the Treasury, 1840–1859, and as such was closely concerned with Irish economic problems at the time of the famine. Clarendon was lord-lieutenant of Ireland, 1847–1852, and his letters are informative on the Treasury orthodoxy of the period, which Trevelyan did much to shape.

2. Biographical information on Dane is taken from his obituary in the London *Times,* February 14, 1940, p. 10.

3. Dane Report, p. 37. The roots of Dane's philosophy of reform may be traced back to Gladstone and the younger William Pitt, who represented the mainstream of British fiscal thinking in the nineteenth century.

4. Dane Report, p. 77.

5. Dane Report, pp. 79, 150.

6. Dane Report, p. 80.

7. The two memoranda by Dane form part of the final report of the royal commission: Appendix B, "Supplementary Note on the History of Opium in India and of the Trade in It with China"; and Appendix C, "Narrative of the Circumstances Preceding and Causes Leading to the First China War." British Parliamentary Papers, Reports from Commissioners, Inspectors and Others, Vol. 42 (1895): *Final Report of the Royal Commission on Opium,* Part II, "Historical Appendices."

8. Dane Report, p. 64. On his tours of inspection of the various salt divisions, Dane was usually accompanied by C. C. Miao (Miao Chiu-chieh), an English-speaking inspectorate officer, later traveling inspector and a leading Chinese member of the inspectorate. Miao, who remained on the mainland after 1949, thus made a considerable contribution to Dane's success. Information supplied to the writer by Mr. Tseng Yang-feng at an interview in Taipei on January 26, 1968.

9. Jordan Papers, Jordan to Langley, January 26, 1914; "A Foreigner Makes Millions for China," *Asia* (July 1917), p. 349.

10. Jordan Papers, Jordan to Langley, January 26, 1914, and Jordan to Langley, February 8, 1914; O. M. Green, *The Foreigner in China* (London, n.d.), p. 119.

11. Dane Report, pp. 113, 88.

12. District Reports, 1918, p. 88.

13. Jordan Papers, Jordan to Langley, February 8, 1914.

14. Dane Report, p. 43.

15. On central government revenue in 1919, see National Archives of the United States, Records of the Bureau of Foreign and Domestic Commerce, Record Group 151, Selected Documents Relating to the Chinese Salt Administration, 1918–1934, File 600, Finance and Investments China, 1919–1929, memorandum of the United States' commercial attaché in Peking, December 31, 1919, "Financial Situation in China."

See also E. L. Woodward and Rohan Butler, eds., *Documents on British Foreign Policy, 1919–1939,* First Series, Vol. VI: *1919* (London, 1956), p. 734, Jordan to Sir J. Tilley, September 24, 1919: "Sir R. Dane and Sir R. Gamble have done wonders in the reorganisation of the Salt Administration which now produces a much larger revenue than the foreign Customs. The Salt and the Customs, both in British hands, not only carry all the foreign obligations, but form almost the only support of the Government which gets practically nothing from the provinces."

16. Dane Report, p. 7.

17. Jordan Papers, Jordan to Langley, February 8, 1914.

18. Dane Report, p. 38.

19. Dane Report, p. 39.

20. The five consortium banks were: the Hongkong and Shanghai Banking Corporation, the Deutsch-Asiatische Bank, the Banque de l'Indo-Chine, the Russo-Asiatic Bank, and the Yokohama Specie Bank. Dane Report, p. 61.

21. For general economic developments in China during Dane's term of office, see *Decennial Reports 1912–1921,* particularly, II, 2, 26–33. For population increase as a factor in the expansion of the salt trade, see below, Chapter 7.

22. District Reports, 1918, p. 76.

23. Dane Report, pp. 35, 39.

24. District Reports, 1918, p. 81; District Reports, 1922, p. 22; District Reports, 1918, p. 8.

25. District Reports, 1913–1917, pp. 96, 99–100; Dane Report, pp. 76–77, 89.

26. Dane Report, pp. 262–268; District Reports, 1918, pp. 74–75; District Reports, 1919–1921, pp. 139, 222.

27. Dane held the two offices of foreign associate chief inspector and foreign adviser in the central salt administration. The first only gave him authority within the inspectorate branch of the salt administration; the second, however, gave him the right to be consulted on all matters within the purview of the whole gabelle. See Dane Report, pp. 46–49.

28. Dane Report, p. 64.

29. Dane Report, p. 39.

30. Dane Report, p. 51.

31. District Reports, 1913–1917, pp. 32, 180–181; Dane Report, pp. 214–217; Customs Report, pp. 282, 315–316, 318–319; China, Ministry of Finance, Chief Inspectorate of the Central Salt Administration, Accounts, 1913–1921, 1922, 1925. An Austrian, A. Wihlfahrt, was the first foreign assistant district inspector at Pakhoi.

32. District Reports,1913–1917, p. 52; Dane Report, p. 118; District Reports, 1919–1921, p. 29; District Reports, 1922, p. 28. The reform of the Su Wu-shu was carried out by a Russian, G. Brauns, as district inspector.

33. Dane Report, p. 88; District Reports, 1913–1917, p. 8; District Reports, 1918, pp. 12–13; District Reports, 1919–1921, pp. 12–15; District Reports, 1922, p. 8. An Englishman, C. G. G. Pearson, was in charge of operations in Shangtung.

34. The Kwangtung farm, for example, was supposed to bring in 5,800,000 taels (Dane Report, p. 126). This compares well with 5,236,800 taels raised by direct administration in Ch'ang-lu in the last year of Hsüan-t'ung. See "Chung-kuo yen-cheng yen-ko shih," p. 10. Ch'ang-lu, slightly larger than Kwangtung, ought to have produced more revenue.

35. Customs Report, pp. 282, 287; Dane Report, pp. 126, 213; Central Salt Administration Accounts; District Reports, 1922, p. 47. The establishment of the Ch'ao-ch'iao assistant inspectorate was the work of a Russian, A. Archangelsky.

36. For example, *CSL:KH,* 512:11–12, and *CSL:KH,* 528:18–19.

37. Dane Report, pp. 40, 39.

38. For Ch'ang-lu and north Anhwei, see Dane Report, p. 156. For Ho-tung, see Dane Report, pp. 157, 91; District Reports, 1913–1917, p. 11. For Jehol and Kalgan, see Dane Report, p. 157. For north Shansi, see District Reports, 1918, p. 3; District Reports, 1919–1921, pp. 174, 255. For the Su Wu-shu, see Dane Report, pp. 222–223. For Fukien, see District Reports, 1918, pp. 45–46; District Reports, 1919–1921, pp. 119–120. For Yunnan, see District Reports, 1913–1917, pp. 194–196; District Reports, 1918, pp. 95–96.

39. For Szechwan, see Dane Report, pp. 175, 238. For Huai-pei, see District Reports, 1913–1917, p. 52; Dane Report, p. 106. For Liang-che, see District Reports, 1918, p. 38; District Reports, 1919–1921, pp. 24, 202; District Reports, 1922, p. 24. For Kansu, see District Reports, 1918, p. 108. For Kwangtung, see District Reports, 1913–1917, pp. 174, 177, 247.

40. District Reports, 1922, p. 33.

41. District Reports, 1918, p. 79.

42. District Reports, 1913–1917, p. 266.

43. District Reports, 1913–1917, p. 86.

44. District Reports, 1913–1917, pp. 241–242.

45. Dane Report, pp. 241, 150, 242.

46. For Ch'ang-lu, see District Reports, 1918, p. 1. For Huai-pei, see District

Reports, 1919–1921, p. 200. For Kansu, see District Reports, 1918, pp. 108–109.

47. Dane Report, p. 187.

48. Dane Report, p. 150.

49. Dane Report, p. 201.

50. Dane Report, p. 38.

51. On Sir Frederick Leith-Ross's mission, see Arthur N. Young, *China's Wartime Finance and Inflation, 1937–1945* (Cambridge, Mass., 1965), p. 7, and Dorothy Borg, *The United States and the Far Eastern Crisis of 1933–1938* (Cambridge, Mass., 1964), Chapter 4.

For the German military mission in the 1930's, see F. F. Liu, *A Military History of Modern China, 1924–1949* (Princeton, 1956), pp. 60–166. For the outlook of Generaloberst von Seeckt, head of the German mission, see his *Gedanken eines Soldaten* (Leipzig, n.d.).

52. On the land tax proposal, see *NCH,* January 16, 1915, p. 153; *NCH,* February 6, 1915, p. 265. See also Jordan Papers, Langley to Jordan, January 20, 1915.

5. Dane's Reforms, 1913–1918: Reform in the Salt Divisions

1. Quoted in *NCH,* April 11, 1914, p. 103.

2. Dane Report, pp. 38, 64.

3. Since these tables form an important part of the evidence for Dane's reforms in the salt divisions, the sources for them must be made clear.

Prereform figures. These are taken, with two exceptions, from Chang Chien's pamphlet, *A Plan for the Reform of the National Salt Administration,* p. 2. Chang Chien states, p. 3, that "The above statistics were collected by the late Manchu Government in 1911," and in the Chinese version (available *YWKMS,* Appendix, pp. 23–75, and *YCTC,* No. 1, *chuan-chien* 1:1–16 [December 1912], and *YCTC,* No. 2, *chuan-chien* 1:17–40 [January 1913]) he states that they are "in accordance with the 1911 records of the former office of the controller-general of the salt gabelle" (*YWKMS,* Appendix, p. 25).

They may, therefore, be regarded as generally reliable, especially as they agree in most cases with the figures given by Dane in his discussion of the state of affairs on his arrival in 1913.

In two cases, however, Huai-nan and Szechwan, Chang Chien's figures need correction if they are not to give a misleading picture of the unreformed gabelle.

Under the heading Liang-huai (*YWKMS,* Appendix, p. 25), Chang Chien gives a figure of 4,896,888 piculs of taxed salt. He obtains this by first multiplying the number of *yin* in Huai-pei and Huai-nan by the appropriate quantification for *yin* in each area at that time, and then by adding the products. In Huai-pei, there were 360,000 *yin* (*YWKMS* actually has the figure 260,000, but this is a typographical error; for the correct figure, see Dane Report, p. 97), which at 4 piculs to the *yin* equals 1,440,000 piculs. In Huai-nan, there were 576,148 *yin,* which at 6 piculs to the *yin* equals 3,456,888 piculs. 1,440,000 plus 3,456,888, equals 4,896,888 piculs.

The Huai-pei computation may be taken as it stands, but in Huai-nan Chang Chien's quantification for the *yin* is open to question, and so is his figure for the number of *yin.*

Chang Chien assumed that the Huai-nan *yin* in 1911 covered only 6 piculs (i.e. 600 catties), though he believed that as much as 100 catties more might be smuggled. According to Dane, however (Dane Report, p. 104), "The size of the Yin in the time of the Ching dynasty was 688 Pa scale catties." Of this amount, 600 catties only were strictly assessed to tax (hence Chang Chien's belief that the *yin* was 6 piculs), the remaining portion being wastage or tare allowance (Dane Report, p. 111). Since the allowances were a legitimate part of the *yin* and passed through the hands of the salt administration, they ought to be included in any realistic figure for taxed salt.

Furthermore, the *pa* scale catty weighed more than the usual *ssu-ma* catty: 1 *pa* scale catty = 1.525 lbs. avoirdupois, while 1 *ssu-ma* catty = 1.400 lbs. avoirdupois (Dane Report, p. 111). The prereform Huai-nan *yin* has therefore been taken at 7.5 piculs instead of Chang Chien's 6. Dane was later to rationalize the whole situation by making the Huai-nan *yin* equivalent to 8 *ssu-ma* piculs (Dane Report, p. 111).

For his figure of 576,148 *yin,* Chang Chien gives no breakdown among the different parts of the division. Other, more detailed accounts of the *yin* system are, therefore, to be preferred. For the local consumption areas of Kiangsu, the so-called *shih-an,* the Customs Report (p. 176) gives 50,848 *yin,* while for the four Yangtze provinces the Dane Report (p. 100) gives an aggregate of 500,000 *yin.* This would give a total number of 550,848 *yin* for Huai-nan, which quantified at 7.5 piculs to the *yin* would produce a figure of 4,131,360 piculs of taxed salt, a considerably higher figure than Chang Chien's. It would seem, however, to be more in accordance with the assessments of Huai-nan of other contemporaries: Chang Hu, for example, the Chinese chief inspector, stated on April 28, 1914 (Dane Report, p. 108), that "The annual quantity of salt consumed in Huai-nan amounts to about 4,000,000 piculs."

For Szechwan, Chang Chien, without basing the figures on any quantification of *yin* and *p'iao,* gives a total of 5,508,600 piculs (*YWKMS,* Appendix, p. 25). That this was indeed the allotted quota is confirmed by District Reports, 1913–1917, p. 68, which indicate that this was composed of 4,125,000 piculs of *yin* or waterborne salt and 1,375,000 piculs of *p'iao* or land-borne salt. There is reason to believe, however, that all of the *yin* figure was never in fact transported. Dane, after detailed examination of the statistics and the units involved (Dane Report, pp. 144–145), gives a figure of 3,732,120 piculs for the *yin* trade. Combined with the *p'iao* figure of 1,375,000 piculs, which is not called in question, this would give a total for Szechwan of 5,107,120 piculs. This figure is confirmed by the allotments made in 1915 when the monopoly system was restored. According to District Reports, 1913–1917, p. 94, "These quantities were based on those transported during the official transportation period," and they amounted to 3,735,980 piculs of *yin* salt and 1,360,928 piculs of *p'iao* salt, making a total of 5,096,908 piculs (Dane Report, p. 176). 5,107,120 piculs has therefore been taken as the prereform figure for taxed salt in Szechwan in preference to Chang Chien's higher figure.

Postreform figures. With one exception, these are all taken from the Central Salt Administration Accounts. The exception is the figure for Kwangtung in 1922: here, because of political disorder, no return at all was made in the central accounts. The inspectors' report, however, contains a figure and this has been used to complete the series. Figures from the inspectors' reports have also been used in the text of the chapter, as they often give a better breakdown within a division than the accounts. The inspectors' total figures, however, often differ slightly from those of the accounts, presumably because of different accounting procedures; these discrepancies will be pointed out whenever they occur.

4. Dane Report, p. 101.

5. Dane Report, p. 102.

6. Dane Report, pp. 102, 107.

7. Dane Report, pp. 106, 113; District Reports, 1918, p. 22.

8. Dane Report, pp. 24, 110, 111, 116.

9. Dane Report, p. 158.

10. Customs Report, p. 176, states that the quota for the *shih-an* was 50,848 *yin*. At 7.5 piculs to the *yin* (see above, note 3), this equals 381,360 piculs.

11. District Reports, 1913–1917, p. 130; District Reports, 1922, p. 16, addition of entries for local consumption areas.

12. The issue of the up-river ports fixed in 1905 and still in force in 1913 was 500,000 *yin* (Dane Report, p. 100). At 7.5 piculs to the *yin* (see above, note 3), this equals 3,750,000 piculs, a figure which was not bettered until 1919.

13. Dane Report, p. 145.

14. Dane Report, p. 176. For taxed salt in Szechwan, see above, note 3.

15. Dane Report, p. 143.

16. District Reports, 1913–1917, p. 66.

17. District Reports, 1913–1917, p. 251.

18. Dane Report, p. 143.

19. In Ch'uan-nan, 1,278,229 piculs of *p'iao* salt were released in 1918 (District Reports, 1918, p. 83), in Ch'uan-pei, the figure was 1,377,148 piculs (District Reports, 1918, p. 92), giving a total for *p'iao* salt for the whole of Szechwan division in that year of 2,655,377 piculs.

By 1918, a certain amount of Ch'uan-pei *p'iao* salt, 384,823 piculs, out of a total of 1,377,148, was being transported, for the first stage of its journey at least, by water. It did not, however, thereby cease to be *p'iao* salt. The essential difference between *yin* and *p'iao* in Szechwan, as in other areas, was one of size: the *yin* covered more salt than the *p'iao,* and it was a matter of fact only that the *p'iao* was used in the main on land routes. For the similar case of Shantung, see above, Chapter 1, note 53.

20. Dane Report, p. 243.

21. Dane Report, pp. 243, 175, 147–148; *Decennial Reports 1902–1911,* I, 272.

22. District Reports, 1918, p. 76; District Reports, 1922, p. 65.

23. The quota for the Szechwan export to Hupei was 1,250,000 piculs, and in 1916, 1,212,500 were actually transported. District Reports, 1918, p. 78; District Reports, 1919–1921, p. 146.

24. District Reports, 1919–1921, p. 3; Dane Report, p. 74.

25. Dane Report, p. 74; District Reports, 1913–1917, p. 119.

26. District Reports, 1913–1917, p. 104; District Reports, 1919–1921, pp. 78–79.

27. Dane Report, p. 163. This increase in tax rate was not general throughout China. Before the reform, Fengtien had enjoyed a particularly low rate of tax, and this increase of rate in 1915 was an attempt to bring it into line with other areas.

28. District Reports, 1913–1917, p. 105; District Reports, 1922, p. 91; District Reports, 1919–1921, p. 6. The Central Salt Administration Accounts give the exact figure for salt issued to the Ki-hei monopoly in 1919 as 1,944,726 piculs.

29. District Reports, 1919–1921, p. 79.

30. "Chung-kuo yen-cheng yen-ko shih," pp. 10–11; Dane Report, p. 36; District Reports, 1922, p. 3.

31. For the Peking-Mukden line, see District Reports, 1918, p. 1; District Reports, 1919–1921, p. 5. For the Chengting–Taiyuan line, see District Reports, 1918, p. 111; District Reports, 1919–1921, pp. 173–174. For the Tientsin–Pukow line, see District Reports, 1919–1921, p. 190. For the Peking–Hankow line, see District Reports, 1913–1917, p. 12; Dane Reports, p. 78; District Reports, 1919–1921, p. 66. For the Lunghai line, see Dane Report, p. 79. For the statement of the Huai-pei inspectors, see District Reports, 1918, p. 26.

32. Dane Report, pp. 180–181, 187.

33. Dane Report, pp. 126, 212; District Reports, 1922, p. 47; *CSL:HT,* 49:15.

34. Chang Chien, p. 2; Central Salt Administration Accounts.

35. Dane Report, p. 136.

36. Dane Report, p. 164. As in Fengtien (see above, note 27), this was a local, not a general, increase of tax rate.

37. District Reports, 1913–1917, p. 80.

38. District Reports, 1918, p. 12; Dane Report, p. 164; District Reports, 1919–1921, p. 13.

39. District Reports, 1922, p. 8.

40. District Reports, 1918, pp. 12–14; District Reports, 1919–1921, p. 104.

41. *Decennial Reports 1912–1921,* I, 209.

42. Dane Report, pp. 156, 194; District Reports, 1913–1917, p. 52; District Reports, 1919–1921, p. 198. For the establishment of the Huai-pei inspectorate at Pan-p'u, see District Reports, 1913–1917, p. 50; Dane Report, p. 105.

43. The figures for prereform Huai-pei have been compiled from the Cus-

toms Report: p. 177 gives 40,413 *yin* for the local consumption areas in northern Kiangsu, which at 440 catties to the *yin* (Dane's quantification, see Dane Report, p. 97) equals 177,817 piculs; p. 144 gives the figures for northern Anhwei and the Ju-kuang districts of Honan.

The figures for salt released in Huai-pei in 1920 are from District Reports, 1919–1921, p. 199. The total for Huai-pei in that year according to the inspectorate reports was 2,917,811, compared to 2,828,563 piculs in the Central Salt Administration Accounts. The difference, which is not significant, is due to different accounting procedure.

44. District Reports, 1913–1917, p. 17, and Dane Report, p. 97, give 149,200 *yin* as the quota for the Huai-pei indirect export to Huai-nan. Although emanating from Huai-pei, these were Huai-nan *yin* (Dane Report, p. 97) and should be quantified at the prereform Huai-nan rate of 7.5 piculs to the *yin* (see above, note 3). 149,200 *yin* at this rate equal 1,119,000 piculs. For the 1919–1921 figures, see District Reports, 1919–1921, pp. 110, 199.

45. District Reports, 1913–1917, pp. 27, 24; Chang Chien, p. 2.

46. District Reports, 1913–1917, p. 52; Dane Report, p. 121; District Reports, 1919–1921, pp. 115–116.

47. District Reports, 1913–1917, pp. 27, 54.

48. District Reports, 1913–1917, p. 228; District Reports, 1918, p. 14; District Reports, 1919–1921, p. 29.

49. District Reports, 1913–1917, p. 86; District Reports, 1919–1921, pp. 29, 202, 204–205; District Reports, 1922, p. 24.

50. *Decennial Reports 1912–1921,* II, 2.

51. Dane Report, p. 260.

52. In 1916, for example, 891,973 piculs were sent to Kwangtung, and 300,334 piculs were sent to Chekiang (District Reports, 1913–1917, p. 149). The Swatow assistant inspectors stated: "The salt consumed by this district is supplied by the works of Chao Chow and of Fukien proportionately in the ratios of 3 and 7" (District Reports, 1913–1917, p. 168), and later, "the whole district of Chao Chiao consumes about 1¼ million piculs of salt a year but the output of the Chao Chiao works is less than half this quantity."

For typhoon damage in Fukien, see District Reports, 1913–1917, p. 237: "In July and September the coastal districts were afflicted by severe typhoons, and at several places, especially at Lienho, the works were laid waste by the waves. The Chaopu works lost about 100,000 piculs and the Shanyan works upwards of 50,000 piculs of salt." See also *CSL:KH,* 477:22.

53. District Reports, 1913–1917, p. 30.

54. District Reports, 1913–1917, pp. 60, 235; Dane Report, p. 123.

55. Dane Report, p. 125.

56. Dane Report, pp. 259–262, 125.

57. District Reports, 1919–1921, p. 208.

58. Dane Report, p. 92.

59. Dane Report, p. 91; District Reports, 1913–1917, pp. 46–47, 82–83, 219.

60. District Reports, 1913–1917, pp. 128, 219.

61. District Reports, 1922, p. 11.

62. District Reports, 1913–1917, p. 82; District Reports, 1919–1921, pp. 106–107.

63. For salt in the northwest, see Dane Report, pp. 157, 159, 256–257. Outside Hua-ting, four main routes were involved: the Fengtien trade to Jehol; the import of Mongolian green salt from the Uchumuch'in lake to both Jehol and Chahar; the Ch'ang-lu trade to the same areas and to northern Shansi: and the import of Mongolian red salt from the Chilantai lake in Alashan into northern Shansi.

64. District Reports, 1918, p. 109.

65. District Reports, 1918, p. 112. The Chin-pei collectors estimated the total salt consumption of their region at 725,000 piculs, of which 410,000 piculs were earth salt. Of the earth salt, only 53,000 piculs paid tax (District Reports, 1919–1921, p. 254). In their first report, the collectors commented on earth salt: "Until the first year of the Republic its manufacture in Chinpei was also forbidden. The local industry is thus a comparatively new one" (District Reports, 1918, p. 112). There is reason to believe, however, that earth salt, whether its manufacture was licit or not, had been a problem for some time in the Taiyuan region at least: see *CSL:HT,* 40:9–11.

66. District Reports, 1913–1917, p. 279; District Reports, 1922, pp. 96, 99.

67. Dane Report, p. 157; District Reports, 1918, p. 108; District Reports, 1919–1921, p. 255.

68. District Reports, 1919–1921, pp. 252, 82; District Reports, 1922, p. 92.

69. District Reports, 1919–1921, p. 170; District Reports, 1922, p. 92.

70. Gale and Ch'en, p. 316; Dane Report, p. 151.

71. Dane Report, p. 153.

72. District Reports, 1913–1917, pp. 194, 271; Dane Report, pp. 168, 207.

73. District Reports, 1919–1921, pp. 154–155, 231.

74. District Reports, 1922, p. 52.

6. Dane and the Salt Interest

1. John K. Fairbank, Edwin O. Reischauer, and Albert M. Craig, *East Asia: The Modern Transformation* (Boston, 1965), p. 132; Robert Laurent, *Les Vignerons de la "Côte d'Or" au XIXe siècle* (Paris, 1958); L. B. Namier, *Avenues of History* (London, 1952), pp. 125–136.

2. J. Le Goff and P. Jeannin, "Questionnaire pour une enquête sur le sel dans l'histoire au moyen âge et aux temps modernes," cyclostyled, Université de Paris, Faculté des Lettres et Sciences Humaines, Section d'Histoire, p. 1. See J. Le Goff, "Une Enquête sur le sel dans l'histoire," *Annales, economies, sociétés, civilisations,* 16.5:959–961 (September–October 1961).

3. Customs Report, pp. 94, 104, 99, 89–90.

4. *Ho-tung yen-fa chih,* Yung-cheng ed. (Taipei, 1966), I, 285–366.

5. Braudel, "Achats et ventes"; Jacques Heers, *Gênes au XVe siècle, activité économique et problèmes sociaux* (Paris, 1961), pp. 349–356; Mauro, pp. 259–277. Also, Aksel E. Christensen, *Dutch Trade to the Baltic about 1600* (Copen-

hagen and The Hague, 1941), pp. 380–381, 401–403. See also Mollat, p. 335, for map of routes.

6. Chaunu, VIII, Part One, 218, 607.

7. For the trade of Setubal, see Mauro, p. 275. For the Dutch salt trade, see D. W. Davies, *A Primer of Dutch Seventeenth Century Overseas Trade* (The Hague, 1961), pp. 15–17; S. G. E. Lythe, *The Economy of Scotland, 1550–1625* (Edinburgh and London, 1960), p. 240; Chaunu, VIII, Part One, 606–612. On salt monopolies in Europe, see Christensen, pp. 106–107; Tawney, pp. 25–26; Mauro, p. 266; Heers, pp. 136–140; M. Marion, *Dictionnaire des institutions de la France au XVIIe et XVIIIe siècles* (Paris, 1923), pp. 247–250, entry "Gabelle."

8. "Le littoral de la Chine est petit par rapport à la superficie du pays et à sa population." J. Crawfurd, *Indian Archipelago* (London, 1820), p. 525; quoted in Louis Dermigny, *La Chine et l'Occident: Le Commerce à Canton au XVIIIe siècle, 1719–1833* (Paris, 1964), I, 48.

For the history of the salt trade in Kweichow, see *Ssu-ch'uan yen-fa chih*, comp. Ting Pao-chen (Chengtu, 1882), chüan 10. See also, Liu Fu, "Pien yen-fa i," *YCTC*, No. 17, *hsüan-lun* 1:17 (February 1915); Wu To, "Ch'uan-yen kuan-yün chih shih-mo," *Chung-kuo chin-tai ching-chi-shih yen-chiu chi-k'an*, 3.2:155 (1935).

9. Marcel Blanchard, "Sel et diplomatie en Savoie et dans les cantons suisses aux XVIIe et XVIIIe siècles," *Annales, economies, sociétés, civilisations*, 15.6:1076–1092 (November–December 1960).

10. Ho Ping-ti, "The Salt Merchants of Yang-chou: A Study of Commercial Capitalism in Eighteenth-Century China," *Harvard Journal of Asiatic Studies*, 17:130–168 (June 1954).

11. Customs Report, p. 6.

12. Mauro, p. 273.

13. Christensen, pp. 389, 112.

14. Lythe, p. 177.

15. Tawney, p. 26.

16. Customs Report, pp. viii, 147–148, 223.

17. TYTP, No. 10, *kung-wen*, 15, 17 (January 1914).

18. Ching Pen-po, "Chiu-ch'ang cheng-shui yü chiu-ch'ang chuan-mai chih pi-chiao," *YCTC*, No. 14, *she-lun* 1:4 (June 1914).

19. Dane Report, p. 252.

20. Kan Kung, "P'i chiu-ch'ang cheng-shui tzu-yu fan-mai chih miu," *TYTP*, No. 1, *hsüan-lun* 3:6 (April 1913).

21. *YCTC*, No. 15, *chi-shih* 1:9 (August 1914).

22. Ching Pen-po, "Chiu-ch'ang cheng-shui," p. 10.

23. *TYTP*, No. 10, *kung-wen*, 15 (January 1914).

24. Liang Ch'i-ch'ao, "Yen-cheng tsa-chih hsü," *YCTC*, No. 1, *hsü*, 4 (December 1912).

25. Ching Pen-po, "Shu Yin-tu yen-wu yen-ko-shih hou," *YCTC*, No. 15, *she-lun* 1:13 (August 1914).

26. *Who's Who in China*, 3rd ed. (Shanghai, 1926), p. 39; *YWKMS*, p. 14; *YCTC*, No. 1, *fa-ling*, 1 (December 1912).

27. Dane Report, p. 183.

28. Dane Report, p. 187.

29. *YWKMS,* p. 15.

30. Ching Pen-po, "Ts'ai-cheng-pu kai-ko yen-wu chi-hua-shu p'ing-lun," *YCTC,* No. 1, *she-lun* 2:7 (December 1912); Dane Report, p. 175.

31. Dane Report, p. 94.

32. *YCTC,* No. 15, *chi-shih* 2:5 (August 1914).

33. *YCTC,* No. 17, *chi-shih* 2:19–21 (February 1915).

34. Dane Report, pp. 147, 173–174; *YCTC,* No. 15, *chi-shih* 2:5 (August 1914).

35. Hsiao K'un, "Yün-nan yen-yün-shih cheng-li yen-wu t'iao-ch'en," *YCTC,* No. 19, *chuan-chien* 1:1–7 (August 1915); Dane Report, pp. 206–207, 168–171; District Reports, 1919–1921, p. 231.

36. Dane Report, pp. 124, 258; District Reports, 1913–1917, p. 236.

37. Dane Report, p. 82.

38. District Reports, 1913–1917, p. 84; District Reports, 1919–1921, p. 143.

39. District Reports, 1922, p. 74; District Reports, 1919–1921, pp. 67, 163.

40. I am grateful to Dr. Edwin G. Beal of the Library of Congress for this bibliographical information.

41. *YCTC,* No. 15, *chi-shih* 1:9–11, 13–14 (August 1914); *YWKMS,* p. 16.

42. Kan Kung, pp. 1–7. The underlying argument was similar to that of T'ao Chu: taxation at source could be applied only where the sources were limited in number. It is surprising that Yunnan was not listed in this category.

43. Kan Kung, p. 6.

44. Hsü Shui-lu, "Lun kai-ko yen-cheng shu," *TYTP,* No. 1, *hsüan-lun* 2:1–7 (April 1913).

45. *TYTP,* No. 10, *kungwen,* 15–18 (January 1914).

46. Dane Report, pp. 82–83; Hui-i, "Ting-en fei-ch'i ch'ang-lu yin-shang chuan tien p'ing-i," *TYTP,* No. 10, *she-lun* 3:1 (January 1914).

47. Dane Report, p. 212; District Reports, 1919–1921, p. 215. Taxation on Kwangtung salt entering Hunan was in fact subsequently reduced. See Dane Report, p. 213.

48. District Reports, 1922, p. 68.

49. *Ssu-ch'uan yen-fa chih,* chüan 17. Two kinds of *yin* circulated in Szechwan: water *yin* (*shui-yin*) and land *yin* (*lu-yin,* also known as *p'iao*). By 1900, all land *yin* represented 8 piculs. Water *yin,* on the other hand, were of two kinds: *hua* or granular salt at 100 piculs each, and *pa* or cake salt at 80 piculs each. Since it is not stated whether water *yin* referred to *hua* or *pa* salt, and since the total quantities of each in Szechwan were approximately equal, all water *yin* have been quantified at the average of 90 piculs a *yin.*

The quota lists for Tzu-liu-ching and Chien-wei in chüan 17 do not contain entries for the *chi-ch'u* trade to Hupei, which was still regarded as irregular. Although Tzu-liu-ching later acquired a monopoly of this trade (District Reports, 1913–1917, p. 67), at the outset both centers participated in it: see *TYTP,* No. 10, *kung-wen,* 16 (January 1914), and *Ssu-ch'uan yen-fa chih,* chüan 11, 12. Tzu-liu-ching, however, was always the more important supplier of the Hupei market (chüan 11), which in the late nineteenth century took 800,000 piculs of Szechwan salt (Customs Report, p. 50). In 1882, therefore, the total turnover at Tzu-liu-ching was probably already at least equal to that at Chien-wei. For the growth of the salt trade of Tzu-liu-ching and the quotas in force on the eve of the revolution of 1911, see *Fu-shun hsien-chih,* 1931 ed. (Taipei, 1967), II, 477–491.

50. Customs Report, p. 57; *TYTP,* No. 10, *kung-wen,* 7–8 (January 1914); Dane Report, p. 241; District Reports, 1913–1917, p. 256.

51. *YCTC,* No. 15, *chi-shih* 2:5–6 (August 1914).

52. *YCTC,* No. 15, *chi-shih* 2:7–8 (August 1914); *YCTC,* No. 17, *chi-shih* 2:21–22 (February 1915).

53. *YCTC,* No. 17, *chi-shih* 2:17–19 (February 1915); Dane Report, p. 177.

54. Dane Report, p. 173.

55. District Reports, 1919–1921, p. 61; District Reports, 1922, pp. 68, 72.

56. Ching Pen-po, "Ts'ai-cheng pu," pp. 6–7.

57. Ching Pen-po, "Chiu-ch'ang cheng-shui," p. 1.

58. Dane Report, pp. 131, 40.

59. Ching Pen-po, "Chiu-ch'ang cheng-shui," pp. 5–6.

60. Ching Pen-po, "Chiu-ch'ang cheng-shui," p. 4.

61. *YCTC,* No. 1, *fu-k'an,* 1 (December 1912); *YCTC,* No. 15, *chi-shih* 1:10–11 (August 1914).

62. *YWKMS,* pp. 6, 10; *YCTC,* No. 15, *chi-shih* 1:10–11 (August 1914).

63. Sir Richard Dane, "Chi-ho tsung-so Ting hui-pan tui-yü Tung-san-sheng yen-wu i-chien shu," *YCTC,* No. 14, *chuan-chien* 1:16 (June 1914); Chang Chien, p. 6; Dane Report, p. 156; Hsü Shui-lu, p. 4.

64. Ching Pen-po, "Ts'ai-cheng pu," p. 2.

65. Ching Pen-po, "Ts'ai-cheng pu," p. 1.

66. Ching Pen-po, "Ts'ai-cheng pu," p. 5; Ching Pen-po, "Shu yin-tu," p. 13.

67. Tseng Yang-feng, *Chung-kuo yen-cheng shih* (Taipei, 1966), p. 9; Chang Chien, pp. 12–14, 18–19.

68. Ching Pen-po, "I-nien-lai yen-cheng kai-ko chih fan-ku," *YCTC,* No. 17, *she-lun* 2:8 (February 1915); Hsü Shui-lu, pp. 3–4; see also *YWKMS,* p. 6.

69. Ching Pen-po, "I-nien-lai," p. 8.

70. *YWKMS,* Appendix, p. 7. The Chekiang program was also printed in the first number of *YCTC.*

71. Ching Pen-po, "Yen-cheng wen-t'i shang-ch'üeh shu," *YCTC,* No. 1, *she-lun* 1:3–5 (December 1912); *YWKMS,* Appendix, p. 63; *YCTC,* No. 1, *t'iao-ch'a* 2: *wai-kuo chih pu* (December 1912).

72. Sun I-shuan, "Salt Taxation in China" (Ph.D. dissertation, University of Wisconsin, 1953), pp. 156–158; Liang Ch'i-ch'ao, p. 6; *YWKMS,* p. 6; Ching Pen-po, "Lun wai-jen kuan-li yen-cheng chih hai," *YCTC,* No. 2, *she-lun* 3:4 (January 1913).

73. Arthur W. Hummel, ed., *Eminent Chinese of the Ch'ing Period, 1644–1912* (Washington, D.C., 1943, 1944), II, 707–708. For the traditional connection between legalism and border problems, see Etienne Balazs, *Chinese*

Civilization and Bureaucracy (New Haven and London, 1964), pp. 198, 206. For the association of high officials of the late empire with border problems, see Louis T. Sigel, "Ch'ing Tibetan Policy (1906–1910)," *Papers on China*, 20:187–189 (Harvard University, East Asian Research Center, 1966).

74. Ching Pen-po, "Lun wai-jen kuan-li," pp. 2, 5; Liang Ch'i-ch'ao, p. 3.

75. Dane Report, p. 6.

76. *YCTC*, No. 15, *chi-shih* 1:12 (August 1914).

77. Ching Pen-po, "Lun wai-jen kuan-li," p. 4; *Report of the Indian Taxation Enquiry Committee, 1924–1925* (Madras, 1926), p. 139; *CSL:HT*, 49:18.

78. *YWKMS*, p. 23; YCTC, No. 14, *chi-shih* 1:8–10 (June 1914).

79. *NCH*, June 6, 1914, p. 751.

80. *YCTC*, No. 15, *chi-shih* 1:13 (August 1914).

81. Ching Pen-po, "Shu yin-tu."

82. Ching Pen-po, "Shu yin-tu," p. 8; Dane Report, pp. 37–39.

83. Dane Report, p. 37.

84. Ching Pen-po, "Shu yin-tu," pp. 12–13.

85. Tseng Yang-feng, *Chung-kuo yen-cheng shih*, p. 286.

86. For example, there was Wang Chen-kan, a writer in the *YCTC* and a member of the Salt Society, who served as chief salt officer at the principal works in Shantung. See *YCTC*, No. 2, *tsa-lu*, 1 (January 1913); *YCTC*, No. 3, *fu-k'an*, 3 (March 1913); *YCTC*, No. 15, *fa-ling*, 4 (August 1914).

87. *NCH*, January 27, 1917, p. 199.

7. The Structure of the Salt Administration in 1920

1. Free trade was in operation in Huai-pei, Fukien, Kwangtung, Yunnan, Szechwan, and the northwest; monopolies continued in Ch'ang-lu, Shantung, Ho-tung, Huai-nan, Sung-kiang and Liang-che; Manchuria was divided, free trade being in operation in Fengtien and monopoly in Kirin and Heilungkiang.

2. *YCTC*, No. 17, *tsa-lu*, 1 (February 1915).

3. *YWKMS*, Appendix, p. 181.

4. Hou Teh-pang, "The Yungli Company: Pioneer of Chemical Industry in China," *China Reconstructs*, 4.3:21–24 (March 1955).

5. Cleveland, p. 66. Nonalimentary salt is here taken to include salt used in fish and meat preservation.

6. By 1922, salt refineries had been established or were projected as follows: at Harbin, one (District Reports, 1913–1917, p. 207); at Newchwang, three (District Reports, 1919–1921, p. 182, and District Reports, 1922, p. 5); in Ch'ang-lu, two (District Reports, 1919–1921, pp. 89, 177); in Shantung, six (District Reports, 1919–1921, p. 187, and District Reports, 1922, p. 9); and in Sung-kiang, one (District Reports, 1919–1921, p. 116).

In addition to these thirteen, there were also seventeen salt refineries which became Chinese by the rendition of Tsingtao at the end of 1922. For the Tsingtao industry, see *Decennial Reports 1912–1921*, I, 222.

7. District Reports, 1919–1921, pp. 107, 176, 188; District Reports, 1922, p. 19; District Reports, 1913–1917, p. 254.

8. The annual breakdown was as follows: 1917, 45,063,144 piculs; 1918, 37,645,178 piculs; 1919, 48,354,855 piculs. These statistics have been compiled from the District Reports, which for most divisions give annual production figures. In some cases, however, no production figures are given: Szechwan, Yunnan, and the northwest for all three years, Kwangtung for 1918 and 1919, Sung-kiang for 1917 and 1919. In these cases, the figure for salt released on payment of tax, as given in the Central Salt Administration Accounts, has been utilized instead. This will tend to understate the total, since, thanks to wastage and smuggling, salt releases were generally below production. Two exceptions have been made in this procedure: for Sung-kiang 1917 and 1919, the 1918 production figure has been assumed constant, since for this division the release figures cannot be used as they include imports from Chekiang; and for Szechwan, Dane's average production figure of 7,000,000 piculs (Dane Report, p. 176) has been assumed constant for all three years as it is probably closer to the truth than the release figures.

9. Tseng Yang-feng, *Chung-kuo yen-cheng shih,* p. 209.

10. District Reports, 1918, p. 22; District Reports, 1919–1921, pp. 1, 89, 176, 94.

11. Hosie, *Manchuria,* p. 152; "Chung-kuo yen-cheng yen-ko shih," p. 9; *Decennial Reports 1902–1911,* I, 209–210.

12. *Decennial Reports 1912–1921,* I, 97; District Reports, 1918, p. 105.

13. "Chung-kuo yen-cheng yen-ko shih," p. 8; Dane Report, p. 266; District Reports, 1913–1917, pp. 103, 276; District Reports, 1919–1921, p. 128. In 1900, of the seven *kuei* or farms into which Kwangtung was divided, the northern *kuei* consumed most salt. See Customs Report, p. 282.

14. District Reports, 1922, pp. 3–4, 47; for Swatow and Pakhoi, see Central Salt Administration Accounts.

15. Timber in a traditional economy is not, of course, a single commodity. Timber for shipbuilding, ordinary building, artisan work, and fuel have each a different demand schedule. For the problem of timber in a traditional economy, see Maurice Lombard, "Un problème cartographié: Le Bois dans la Méditerranée musulmane (VIIe–XIe siècles)," *Annales, economies, sociétés, civilisations,* 14.2:234–254 (April–June 1959).

16. *Ch'ang-lu yen-fa chih,* Yung-cheng ed. (Taipei, 1966), I, 627–634; *Ho-tung yen-fa chih,* I, 216–220.

17. District Reports, 1919–1921, pp. 20, 110, 17; District Reports, 1913–1917, p. 40.

18. Mollat, p. 128. For the salt trade in nineteenth-century France, see Mollat, pp. 259–262. For the Indian salt trade in the early twentieth century, see Indian Tariff Board, *Evidence Recorded During Enquiry on the Salt Industry* (Calcutta, 1930).

19. *Decennial Reports 1912–1921,* I, 90; District Reports, 1919–1921, p. 94. For the predominant role of millet in Manchurian trade and its connection with the movement of salt, see Hosie, *Manchuria,* pp. 240, 239.

20. By the early 1930's, 60 to 70 percent of the export of bean products was going to Europe, especially Germany; see *The Manchoukuo Yearbook, 1934* (Tokyo, 1934), p. 269. The soybean industry in Manchuria was thus partly a function of the chemical industry in Europe. For the trade inside China, see Hosie, *Manchuria,* p. 244; *Decennial Reports 1902–1911,* II, 118; *Decennial Reports 1912–1921,* II, 171, 175–176. For Ch'ang-lu, see *Decennial Reports 1912–1921,* I, 154–157, 161.

21. *Decennial Reports 1902–1911,* I, 135. For the contrasting growth rates of beans, kaoliang, and millet production in Manchuria, see *The Manchoukuo Yearbook, 1934,* p. 263.

22. *Ho-tung yen-fa chih,* I, 285, 313.

23. Dane Report, p. 143; District Reports, 1919–1921, p. 144; Customs Report, p. 141.

24. District Reports, 1919–1921, p. 200.

25. District Reports, 1913–1917, p. 40; District Reports, 1918, p. 1; District Reports, 1919–1921, p. 109.

26. District Reports, 1913–1917, p. 205; *Decennial Reports 1902–1911,* I, 359.

27. District Reports, 1919–1921, pp. 238, 159.

28. Assessment and collection, the two main processes of tax-levying, are susceptible to further analysis; speed of transfer from original collection to ultimate consolidated receipt is another major criterion of modernization. For the European model, see J. E. D. Binney, *British Public Finance and Administration, 1774–92* (Oxford, 1958).

29. Dane Report, pp. 24–27; District Reports, 1913–1917, pp. 86, 220, 230; District Reports, 1918, p. 44.

30. Customs Report, p. 177; 50,848 Huai-nan *yin,* at 7.5 piculs to the *yin,* equal 381,360 piculs. Dane Report, p. 118; District Reports, 1918, pp. 20, 43.

31. Customs Report, p. 223. The Nanking customs commissioner, who was generally well informed, gave a figure of 6,958,900 taels for the *cheng-k'o* of the whole empire, while the Foochow commissioner (Customs Report, p. 249) suggested 5,730,000 taels. The higher figure has been accepted here in order to undervalue rather than exaggerate Dane's achievement.

32. *YWKMS,* Appendix, p. 180.

33. Jordan Papers, Jordan to Langley, May 29, 1918; Jordan Papers, Jordan to Sir Harcourt Butler, April 11, 1916; Dane Report, pp. 62–64; Sir Charles Addis, "The Finance of China," *Edinburgh Review,* 230.470:299 (October 1919).

34. For forced loans in Hunan, see District Reports, 1919–1921, pp. 162–163. For the looting of the Bank of Hunan in 1918, see Dane Report, p. 251; for the intimidation of the Chinese district inspector in Fukien, see District Reports, 1922, p. 36. For an illegal surtax by Wu P'ei-fu in Hupei in 1921, see District Reports, 1919–1921, p. 238.

35. District Reports, 1919–1921, pp. 126–128, 220.

36. Dane Report, pp. 254–256. For an example of this system in operation, in Ch'uan-nan salt division in 1919, see District Reports, 1919–1921, p. 60.

37. Dane Report, pp. 154, 167–168. In 1915, total revenue in Yunnan came to $3,021,541; District Reports, 1913–1917, p. 101.

38. District Reports, 1913–1917, p. 199.

39. Dane Report, pp. 236–237, 247.

40. Dane Report, pp. 248, 253–254. For the diplomatic representations of the consortium powers against the Canton government's actions in 1918, see China, Wai-chiao-pu (Ministry of Foreign Affairs) Archives, Salt Administration, 1912–1926, *han* 2 file dated June–October 1918.

41. District Reports, 1918, p. 70.

42. District Reports, 1919–1921, p. 217.

43. District Reports, 1919–1921, p. 215.

44. District Reports, 1919–1921, p. 219; District Reports, 1922, p. 72.

45. Dane Report, pp. 16, 274–275; Jordan Papers, Jordan to Langley, February 10, 1916.

46. Jordan Papers, Dane to Addis, undated, enclosed in Jordan to Langley, April 16, 1916. It is not certain that this letter was ever sent, though Dane and Chang Hu addressed a similar one to the banks' representatives on May 11, 1916. Dane Report, pp. 233–235.

47. Jordan Papers, Max Muller to Jordan, April 4, 1919.

48. National Archives of the United States, Records of the Bureau of Foreign and Domestic Commerce, Record Group 151, Selected Documents Relating to the Chinese Salt Administration, 1918–1934, File 600, Finance and Investments China, 1919–1929, Frank Rhea, Trade Commissioner, to F. R. Eldridge, Chief Far Eastern Division, Bureau of Foreign and Domestic Commerce, Washington, April 29, 1922, August 3, 1922, October 9, 1922.

The quotation is from a cutting from the *Peking and Tientsin Times,* October 4, 1922, enclosed in the letter of October 9, 1922.

49. Jordan Papers, Jordan to Langley, June 13, 1916.

50. Young, pp. 39–50. Also Nicholas R. Clifford, "Sir Frederick Maze and the Chinese Maritime Customs, 1937–1941," *Journal of Modern History,* 37.1:18–34 (March 1965).

51. Dane Report, p. 15; *YCTC,* No. 19, *fa-ling,* 6 (August 1915); District Reports, 1919–1921, pp. 97, 105.

52. District Reports, 1919–1921, pp. 55, 227.

53. For the situation at Ichang in 1922, see District Reports, 1922, pp. 73–75. For Wu P'ei-fu's title, see District Reports, 1922, p. 77. For other interferences by Wu P'ei-fu with the salt administration in 1922, see China, Wai-chiao-pu (Ministry of Foreign Affairs) Archives, Salt Administration, 1912–1926, *han* 3 file dated February 1922.

54. Jordan Papers, Jordan to Langley, September 17, 1914.

55. Information supplied to the writer by Mr. Tseng Yang-feng at an interview in Taipei on January 26, 1968.

56. Rohan Butler and J. P. T. Bury, eds., *Documents on British Foreign Policy, 1919–1939,* First Series, Vol. XIV: *Far Eastern Affairs, April 1920–February 1922* (London, 1966), p. 430, Memorandum by Sir J. Jordan Respecting American Suggestions for Agenda for the Conference, October 17, 1921.

57. For Ma Yin-ch'u's views on salt reform in 1928, see the program drawn up by him and by Ching Pen-po's old associate, Chuang Sung-fu, *YWKMS,* Appendix, pp. 190–194. For Ma Yin-ch'u's later career, see K. R. Walker, "A Chinese Discussion on Planning for Balanced Growth: A Summary of the Views of Ma Yin-ch'u and His Critics," in C. D. Cowan, ed., *The Economic Development of China and Japan* (London, 1964), pp. 160–161.

8. Synarchy

1. Fernand Braudel, *La Méditerranée et le monde méditerranéen à l'époque de Philippe II* (Paris, 1949), pp. 531, 1035, 600, 127.
2. Braudel, *La Méditerranée,* p. 599.
3. Braudel, *La Méditerranée,* p. 600.
4. Stanford J. Shaw, *Ottoman Egypt in the Age of the French Revolution* (Cambridge, Mass., 1964).
5. Vernon J. Puryear, *Napoleon and the Dardanelles* (Berkeley and Los Angeles, 1951).
6. British Parliamentary Papers, *Papers Relating to Administrative and Financial Reforms in Turkey, 1858–61* (London, 1861), Bulwer to Lord John Russell, November 16, 1860.
7. British Parliamentary Papers, *Papers Relating to Administrative and Financial Reforms in Turkey, 1858–61,* Bulwer to Russell, April 24, 1860.
8. Persia, Fifth Quarterly Report of the Administrator General of the Finances of Persia, December 23, 1923.
9. W. Morgan Shuster, *The Strangling of Persia* (New York, 1912), pp. 49–50.
10. Shuster, pp. 287–288.
11. Persia, Ninth Quarterly Report of the Administrator General of the Finances of Persia, December 21, 1924.
12. Persia, Fifth Quarterly Report of the Administrator General of the Finances of Persia, December 23, 1923.
13. Ottoman Public Debt, Special Report 1909–1910, p. 30.
14. Ottoman Public Debt, Special Report 1909–1910.
15. Shuster, p. 330.
16. The Jesuit accommodation policy should be understood against the background of the sixteenth-century debate over grace and nature. The Thomist view of nature was reasserted by the Dominican Vitoria and the Jesuit Suarez, both of whom applied it to problems of international relations.
17. Tseng Yang-feng, "I Ting-en," *Yen-yeh t'ung-hsün,* 77:10–11 (January 1958). Tseng puts Dane on a level with Kuan-tzu and Liu Yen.
18. British Foreign Office Records, F.O. 405/233, F. T. A. Ashton-Gwatkin, Memorandum Respecting Foreign Control in Chinese Government Services, October 10, 1921.
19. Sir Telford Waugh, *Turkey, Yesterday, Today and Tomorrow* (London, 1930), p. 157.
20. For Japan, see Sun I-shuan, pp. 156–159; for Wieliczka, see Michel

Kincler, *L'Industrie du sel en Pologne* (Nancy, 1935); and for Brazil, see Dioclecio D. Duarte, *A Indústria extrativa do sal e a sua importancia na economia do Brasil* (Rio de Janeiro, 1941).

21. Information supplied to the writer by Mr. Tseng Yang-feng at an interview in Taipei on January 26, 1968.

22. Hou Chi-ming, *Foreign Investment and Economic Development in China, 1840–1936* (Cambridge, Mass., 1965), pp. 130–138.

23. British Foreign Office Records, F.O. 405/233, B. C. Newton, Memorandum Respecting the Unification of China's Railways, October 10, 1921.

24. British Foreign Office Records, F.O. 405/233, Memorandum Respecting Currency Reform in China, October 10, 1921.

25. British Foreign Office Records, F.O. 405/233, Sir Victor Wellesley, Memorandum Respecting Intervention in China, October 10, 1921.

26. British Parliamentary Papers, *Papers Relating to Administrative and Financial Reforms in Turkey, 1858–61,* Bulwer to Russell, November 12, 1860.

27. Christopher Hill, *The Century of Revolution, 1603–1714* (Edinburgh, 1961), p. 141. For the Hargrave manuscripts reform project, see G. E. Aylmer, *The King's Servants* (London, 1961), pp. 234–237.

28. H. R. Trevor-Roper, *The Crisis of the Seventeenth Century* (New York and Evanston, 1968).

29. This generalization is intended to refer primarily to the Ottoman empire, Persia, India, and China, but it would seem applicable to Japan, though it might be questioned whether even the late Tokugawa state should be termed a bureaucratic monarchy.

30. Shuster, p. 333.

31. British Parliamentary Papers, *Papers Relating to Administrative and Financial Reforms in Turkey, 1858–61,* Bulwer to Russell, November 12, 1860.

32. Donald C. Blaisdell, *European Financial Control in the Ottoman Empire* (New York, 1929), p. 124.

33. Persia, Twelfth Quarterly Report of the Administrator General of the Finances of Persia, September 22, 1925.

34. British Foreign Office Records, F.O. 228/3035, Jordan to Foreign Office, June 7, 1919.

35. Young, p. 53.

Bibliography

Addis, Sir Charles. "The Finance of China," *Edinburgh Review*, vol. 230, no. 470 (October 1919).

Aylmer, G. E. *The King's Servants*. London: Routledge and Kegan Paul, 1961.

Balazs, Etienne. *Chinese Civilization and Bureaucracy*. New Haven and London: Yale University Press, 1964.

Binney, J. E. D. *British Public Finance and Administration, 1774–92*. Oxford: Clarendon Press, 1958.

Blaisdell, Donald C. *European Financial Control in the Ottoman Empire*. New York: Columbia University Press, 1929.

Blanchard, Marcel. "Sel et diplomatie en Savoie et dans les cantons suisses aux XVIIe et XVIIIe siècles," *Annales, economies, sociétés, civilisations*, 15.6:1076–1092 (November–December 1960).

Bland, J. O. P., and E. Backhouse. *China under the Empress Dowager*. London: William Heinemann, 1912.

Borg, Dorothy. *The United States and the Far Eastern Crisis of 1933–1938*. Cambridge, Mass.: Harvard University Press, 1964

Braudel, Fernand. *La Méditerranée et le monde méditerranéen à l'époque de Philippe II*. Paris: Librairie Armand Colin, 1949.

——— "Achats et ventes de sel à Venise (1587–1593)," *Annales, economies, sociétés, civilisations*, 16.5:961–965 (September–October 1961).

Bridbury, A. R. *England and the Salt Trade in the Later Middle Ages*. Oxford: Clarendon Press, 1955.

British Foreign Office Records, Public Record Office, London.

British Parliamentary Papers, *Papers Relating to Administrative and Financial Reforms in Turkey, 1858–61*. London, 1861.

British Parliamentary Papers, Reports from Commissioners, Inspectors and Others, Vol. 42 (1895): *Final Report of the Royal Commission on Opium*.

British Parliamentary Papers, China No. 1 (1908), *Correspondence Respecting the Opium Question in China*.

British Parliamentary Papers, China No. 3 (1909), *Reports Respecting the Opium Question in China*.

British Parliamentary Papers, China No. 1 (1912), *Correspondence Respecting the Affairs of China.*

British Parliamentary Papers, China No. 3 (1912), *Further Correspondence Respecting the Affairs of China.*

Brunnert, H. S., and V. V. Hagelstrom. *Present-Day Political Organization of China*, tr. A. Beltchenko and E. E. Moran. Shanghai: Kelly and Walsh, 1912.

Butler, Rohan, and J. P. T. Bury, eds. *Documents on British Foreign Policy, 1919–1939.* First Series. Vol. XIV: *Far Eastern Affairs, April 1920–February 1922.* London: Her Majesty's Stationery Office, 1966.

Carlson, Ellsworth. "The Kailan Mines, 1879–1912," *Papers on China*, 3:24–77. Harvard University, East Asian Research Center, 1949.

Chang Chien. *A Plan for the Reform of the National Salt Administration.* Shanghai: The National Review Office, 1913.

Chang Kia-ngau. *China's Struggle for Railroad Development.* New York: John Day, 1943.

Ch'ang-lu yen-fa chih 長蘆鹽法志 (Treatise on the Ch'ang-lu salt laws). Yung-cheng ed. 2 vols. Taipei: T'ai-wan hsüeh-sheng shu-chü, 1966.

Chaunu, Huguette and Pierre. *Séville et l'Atlantique* (*1504–1650*). 8 vols. Paris: S.E.V. P.E.N., 1955–1959.

Ch'en, Jerome. *Yuan Shih-k'ai.* London: George Allen and Unwin, 1961.

China, Imperial Maritime Customs. *Decennial Reports 1882–1891.* Shanghai, 1893.

——— *Decennial Reports 1902–1911.* 3 vols. Shanghai: The Statistical Department of the Inspectorate General of Customs, 1912.

——— *Salt: Production and Taxation.* V Office Series, Customs Papers, No. 81. Shanghai: The Statistical Department of the Inspectorate General of Customs, 1906.

China, Maritime Customs. *Decennial Reports 1912–1921.* 2 vols. Shanghai: The Statistical Department of the Inspectorate General of Customs, 1922.

China, Ministry of Finance, Chief Inspectorate of the Central Salt Administration. Accounts, 1913–1921, 1922, 1925.

China, Wai-chiao-pu 外交部 (Ministry of Foreign Affairs) Archives. Salt Administration, 1912–1926. 45 ts'e.

The China Yearbook, 1945. China Publishing Co.

Ching Pen-po 景本白. "Yen-cheng wen-t'i shang-ch'üeh shu" 鹽政問題商榷書 (Discussions on the problems of the salt administration), *Yen-cheng tsa-chih*, No. 1, *she-lun* 1:1–7 (December 1912).

——— "Ts'ai-cheng-pu kai-ko yen-wu chi-hua-shu p'ing-lun" 財政部改革鹽務計劃書評論 (A critical discussion of the ministry of finance's plan for the reform of the salt administration), *Yen-cheng tsa-chih*, No. 1, *she-lun* 2:1–21 (December 1912).

——— "Yen-cheng wen-t'i shang-ch'üeh shu" 鹽政問題商榷書 (Discussions on the problems of the salt administration, continuation), *Yen-cheng tsa-chih*, No. 2, *she-lun* 1:1–8 (January 1913).

——— "Lun wai-jen kuan-li yen-cheng chih hai" 論外人管理鹽政之害 (On the evils of foreign control of the salt administration), *Yen-cheng tsa-chih*, No. 2, *she-lun* 3:1–6 (January 1913).

——— "Chiu-ch'ang cheng-shui yü chiu-ch'ang chuan-mai chih pi-chiao" 就場徵稅與就場專賣之比較 (Comparison of the taxation-at-source and monopoly systems), *Yen-cheng tsa-chih*, No. 14, *she-lun* 1:1–10 (June 1914).

——— "Shu Yin-tu yen-wu yen-ko-shih hou" 書印度鹽務沿革史後 (Postscript to the history of the Indian salt administration), *Yen-cheng tsa-chih*, No. 15, *she-lun* 1:1–14 (August 1914).

——— "I-nien-lai yen-cheng kai-ko chih fan-ku" 一年來鹽政改革之返顧 (A review of the past year's reforms in the salt administration), *Yen-cheng tsa-chih*, No. 17, *she-lun* 2:1–10 (February 1915).

——— "Lun pao-hsiao" 論報効 (On contributions), *Yen-cheng tsa-chih*, No. 19, *she-lun* 1:1–6 (August 1915).

——— *Yen-wu ko-ming shih* 鹽務革命史 (A history of the revolution in the salt administration). Nanking: Ching-yen tsung-hui yen-cheng tsa-chih she, 1929.

Christensen, Aksel E. *Dutch Trade to the Baltic about 1600*. Copenhagen and The Hague: Einar Munksgaard and Martinus Nijhoff, 1941.

Chu, Samuel C. *Reformer in Modern China: Chang Chien, 1853–1926*. New York and London: Columbia University Press, 1965.

"Chung-kuo yen-cheng yen-ko shih" 中國鹽政沿革史 (A history of the development of the Chinese salt administration), *Yen-cheng tsa-chih*, No. 19, *chuan-chien* 2:1–16 (August 1915).

Clarendon Papers, Bodleian Library, Oxford.

Cleveland, F. A. "Statistical Review of the Work of the Inspectorate, 1913–1933, with Special Attention Given to the Evaluation of Results Achieved During the Last Five Years." Chinese National Government, Ministry of Finance, Inspectorate of Salt Revenue, n.d.

Clifford, Nicholas R. "Sir Frederick Maze and the Chinese Maritime Customs, 1937–1941," *Journal of Modern History*, 37.1:18–34 (March 1965).

Clow, A. and N. L. *The Chemical Revolution*. London: Batchworth Press, 1952.

——— "The Chemical Industry: Interaction with the Industrial Revolution," in Charles Singer, E. J. Holmyard, A. R. Hall, and Trevor I. Williams, eds., *A History of Technology*, IV, 230–257. 5 vols. Oxford: Clarendon Press, 1954–1958.

Dane, Sir Richard. "Chi-ho tsung-so Ting hui-pan tui-yü Tung-san-sheng yen-wu i-chien shu" 稽核總所丁會辦對於東三省鹽務意見書 (Report of associate chief inspector Dane on the Manchurian salt administration), *Yen-cheng tsa-chih*, No. 14, *chuan-chien* 1:1–19 (June 1914).

Davies, D. W. *A Primer of Dutch Seventeenth Century Overseas Trade*. The Hague: Martinus Nijhoff, 1961.

Davidson, Robert J., and Isaac Mason. *Life in West China*. London: Headley Brothers, 1905.

Dawson, Owen L. "China's Cotton Outlook," *The China Mainland Review*, 2.3:174–184 (December 1966).

Delafosse, M., and C. Laveau. *Le Commerce du sel de Brouage au XVIIe et XVIIIe siècles*. Paris: Librairie Armand Colin, 1961.

Dermigny, Louis. *La Chine et l'Occident: Le Commerce à Canton au XVIIIe siècle*. 3 vols. Paris: S.E.V.P.E.N., 1964.

Dictionary of National Biography, 1912–1921. Oxford: Oxford University Press, 1927.

Duarte, Dioclecio D. *A Indústria extrativa do sal e a sua importancia na economia do Brasil*. Rio de Janeiro, 1941.

Eisenstadt, S. N. *The Political Systems of Empires*. London: The Free Press of Glencoe, 1963.

Encyclopaedia Britannica. London: Encyclopaedia Britannica Ltd., 1957. 24 vols.

Fairbank, John K., Edwin O. Reischauer, and Albert M. Craig. *East Asia: The Modern Transformation*. Boston: Houghton Mifflin Company, 1965.

Fairbank, John K. and Ssu-yü Teng. *Ch'ing Administration: Three Studies*. Cambridge, Mass.: Harvard University Press, 1960.

Feuerwerker, Albert. *China's Early Industrialisation*. Cambridge, Mass.: Harvard University Press, 1958.

Finer, S. E. *The Life and Times of Sir Edwin Chadwick*. London: Methuen and Co., Ltd., 1952.

First Five-Year Plan for Development of the National Economy of the People's Republic of China in 1953–1957. Peking: Foreign Languages Press, 1956.

Fu-shun hsien-chih 富順縣志 (Fu-shun district gazetteer). 1931 ed., 4 vols. Taipei: T'ai-wan hsüeh-sheng shu-chü, 1967.

"La Gabelle du sel," *La Chine*, 39:347–362, 40:431–446 (April 1923).

Gale, Esson M., and Ch'en Sung-ch'iao. "China's Salt Administration: Excerpts from Native Sources," *Journal of Asiatic Studies*, 2.1:273–316 (June 1959).

Gamble, Sidney D. *Peking: A Social Survey*. New York: George H. Doran Company, 1921.

Gestrin, Ferdo. "Economie et société en Slovénie au XVIe siècle," *Annales, economies, sociétés, civilisations*, 17.4:663–690 (July-August 1962).

Goodwin, A., ed. *The New Cambridge Modern History*, Vol. VIII: *The American and French Revolutions, 1763–93*. Cambridge, England: Cambridge University Press, 1965.

Green, O. M. *The Foreigner in China*. London: Hutchinson and Co., Limited, n.d.

——— *The Story of China's Revolution*. London: Hutchinson and Co., Limited, n.d.

Heers, Jacques. *Gênes au XVe siècle, activité économique et problèmes sociaux*. Paris: S.E.V.P.E.N. 1961.

Hill, Christopher. *The Century of Revolution, 1603–1714*. Edinburgh: Thomas Nelson and Sons Ltd., 1961.

Ho Ping-ti. "The Salt Merchants of Yang-chou: A Study of Commercial Capitalism in Eighteenth-Century China," *Harvard Journal of Asiatic Studies*, 17:130–168 (June 1954).

——— *Studies on the Population of China, 1368–1953*. Cambridge, Mass.: Harvard University Press, 1959.

Hosie, Alexander. *Three Years in Western China*. London, 1890.

——— *Manchuria: Its People, Resources and Recent History*. London: Methuen and Co., 1901.

——— "The Salt Production and Salt Revenue of China," *Nineteenth Century*, 447:1119–1143 (May 1914).

Ho-tung yen-fa chih 河東鹽法志 (Treatise on the Ho-tung salt laws). Yung-cheng ed. 2 vols. Taipei: T'ai-wan hsüeh-sheng shu-chü, 1966.

Hou Chi-ming. *Foreign Investment and Economic Development in China, 1840–1937*. Cambridge, Mass.: Harvard University Press, 1965.

Hou Teh-pang. "The Yungli Company: Pioneer of Chemical Industry in China," *China Reconstructs*, 4. 3:21–24 (March 1955).

Hsiao K'un 蕭堃. "Yün-nan yen-yün-shih cheng-li yen-wu t'iao-ch'en" 雲南鹽運使整理鹽務條陳 (The Yunnan salt commissioner's proposals for the reorganization of the salt administration), *Yen-cheng tsa-chih*, No. 19, *chuan-chien* 1:1–7 (August 1915).

Hsü Shui-lu 許蜕廬. "Lun kai-ko yen-cheng shu" 論改革鹽政書 (Discussing the reform of the salt administration), *T'an-yen ts'ung-pao*, No. 1, *hsüan-lun* 2:1–7 (April 1913).

Hughes, Edward. *Studies in Administration and Finance, 1558–1825*. Manchester: Manchester University Press, 1934.

Hui-i 悔逸. "Ting-en fei-ch'i ch'ang-lu yin-shang chuan tien p'ing-i" 丁恩廢棄長蘆引商專電平義 (The justice of Dane's telegram abolishing the monopoly of the Ch'ang-lu *yin* merchants), *T'an-yen ts'ung-pao*, No. 10, *she-lun* 3:1–6 (January 1914).

Hummel, Arthur W., ed., *Eminent Chinese of the Ch'ing Period, 1644–1912*. 2 vols. Washington, D.C.: United States Government Printing Office, 1943, 1944.

Indian Tariff Board. *Evidence Recorded During Enquiry on the Salt Industry*. 2 vols. Calcutta, 1930.

Johnston, R. F. *Twilight in the Forbidden City*. London: Victor Gollancz Ltd., 1934.

Jordan, W. K. *The Charities of London*. London: George Allen and Unwin Ltd., 1960.

Jordan Papers, Public Record Office, London.

Kan Kung 干公. "P'i chiu-ch'ang cheng-shui tzu-yu fan-mai chih miu" 闢就場徵稅自由販賣之謬 (Exposing the errors of the taxation-at-source and free trade system), *T'an-yen ts'ung-pao*, No. 1, *hsüan-lun* 3:1–7 (April 1913).

Kent, P. H. *The Passing of the Manchus*. London: E. Arnold, 1912.

Kincler, Michel. *L'Industrie du sel en Pologne*. Nancy, 1935.

Laurent, Robert. *Les Vignerons de la "Côte d'Or" au XIXe siècle*. 2 vols. Paris: Société *Les Belles Lettres*, 1958.

Le Goff, J. "Une Enquête sur le sel dans l'histoire," *Annales, economies, sociétés, civilisations*, 16.5:959–961 (September-October 1961).

Le Goff, J., and P. Jeannin. "Questionnaire pour une enquête sur le sel dans l'histoire au moyen âge et aux temps modernes." Cyclostyled, Université de Paris, Faculté des Lettres et Sciences Humaines, Section d'Histoire.

Li Chien-nung. *The Political History of China, 1840–1928*. Princeton: D. Van Nostrand Company, Inc., 1956.

Li Jo-ping, ed. *A Collection of Modern Chinese Essays*.

Liang Ch'i-ch'ao 梁啟超. "Yen-cheng tsa-chih hsü 鹽政雜誌序 (Preface to the salt administration magazine), *Yen-cheng tsa-chih*, No. 1, *hsü* 1–7 (December 1912).

Liu, F. F. *A Military History of Modern China, 1924–1949*. Princeton: Princeton University Press, 1956.

Liu Fu 劉馥. "Pien yen-fa i" 變鹽法議 (On changing the salt laws), *Yen-cheng tsa-chih*, No. 17, *hsüan-lun* 1:1–26 (February 1915).

Liu, James T. C. *Reform in Sung China*. Cambridge, Mass.: Harvard University Press, 1959.

Liu, Kwang-ching. *Anglo-American Steamship Rivalry in China, 1862–1874*. Cambridge, Mass.: Harvard University Press, 1962.

Lombard, Maurice. "Un Problème cartographié: Le Bois dans la Méditerranée musulmane (VIIe–XIe siècles)," *Annales, economies, sociétés, civilisations*, 14.2: 234–254 (April–June 1959).

Lythe, S. G. E. *The Economy of Scotland, 1550–1625*. Edinburgh and London: Oliver and Boyd, 1960.

The Manchoukuo Yearbook, 1934. Tokyo: Toa-Keizai Chosakyoku, 1934.

Marion, M. *Dictionnaire des institutions de la France au XVIIe et XVIIIe siècles*. Paris: Auguste Picard, 1923.

Mauro, Frédéric. *Le Portugal et l'Atlantique au XVIIe siècle (1570–1670)*. Paris: S.E.V.P.E.N., 1960

Metzger, Thomas A. "T'ao Chu's Reform of the Huai-pei Salt Monopoly," *Papers on China*, 16:1–39. Harvard University, East Asian Research Center, 1962.

——— "Command Economy and Market Economy in Ming and Ch'ing Times." Unpublished paper.

Mollat, Michel, ed. *Le Rôle du sel dans l'histoire*. Paris: Presses Universitaires de France, 1968.

Morley, John. *The Life of William Ewart Gladstone*. 3 vols. London: Macmillan and Co., Limited, 1903.

Morse, H. B. *The Trade and Administration of the Chinese Empire*. London: Longmans, Green, and Co., 1908.

——— *The International Relations of the Chinese Empire*. 3 vols. London: Longmans, Green, and Co., 1910–1918.

Namier, L. B. *Avenues of History*. London: Hamish Hamilton, 1952.

National Archives of the United States, Records of the Bureau of Foreign and Domestic Commerce, Record Group 151, Selected Documents Relating to the Chinese Salt Administration, 1918–1934.

Needham, Joseph. *Science and Civilisation in China*. 7 vols. Cambridge, England: Cambridge University Press, 1954– .

Nef, J. U. *The Rise of the British Coal Industry*. 2 vols. London: George Routledge, 1932.

North China Herald and Supreme Court and Consular Gazette. 1911–1918.

Ottoman Public Debt. Annual and Special Reports, 1904–05 to 1907–08, 1909–10 to 1913–14, 1919–20, 1922–23.

Parker, E. H. *China: Her History, Diplomacy and Commerce*. London: John Murray, 1917.

Peck, Graham. *Through China's Wall*. London: Collins, 1941.

Persia. Quarterly Reports of the Administrator General of the Finances of Persia. 1923–1928.

Pliny the elder, *Naturalis Historia*: William Heinemann Ltd. and Harvard University Press, 1963.

Polo, Marco. *The Travels of Marco Polo*, tr. R. E. Latham. Penguin Books ed. Harmondsworth: Penguin Books Ltd., 1958.

Powell, Ralph L. *The Rise of Chinese Military Power, 1895–1912*. Princeton: Princeton University Press, 1955.

P'u Yu-shu. "The Consortium Reorganisation Loan to China, 1911–1914: An Episode in Pre-war Diplomacy and International Finance." Ph.D. dissertation, University of Michigan, 1951.

Pugh, R. B. "The Colonial Office, 1801–1925," in E. A. Benians, James Butler, and C. E. Carrington, eds., *The Cambridge History of the British Empire*, III, 711–768. Cambridge, England: Cambridge University Press, 1959.

Puryear, Vernon J. *Napoleon and the Dardanelles*. Berkeley and Los Angeles: University of California Press, 1951.

Report by Sir Richard Dane, K.C.I.E., on the Reorganisation of the Salt Revenue Administration in China, 1913–1917.

Reports by the District Inspectors, Auditors and Collectors on the Reorganisation of the Salt Revenue Administration in China, 1913–1917, 1918, 1919–1921, 1922.

Report of the Indian Taxation Enquiry Committee, 1924–1925. Madras, 1926.

Richthofen, Ferdinand von. *Baron Richthofen's Letters, 1870–1872*. 2nd ed. Shanghai: North China Herald Office, 1903.

Rostow, W. W. *The Stages of Economic Growth*. Cambridge, England: Cambridge University Press, 1962.

Schurmann, Herbert Franz. *Economic Structure of the Yüan Dynasty*. Cambridge, Mass.: Harvard University Press, 1956.

——— *Ideology and Organization in Communist China*. Berkeley and Los Angeles: University of California Press, 1966.

Schwartz, Benjamin. *In Search of Wealth and Power: Yen Fu and the West*. Cambridge, Mass.: The Belknap Press of Harvard University Press, 1964.

Sée, Henri. *Histoire économique de la France*. Paris: Librairie Armand Colin, 1948.

Seeckt, Generaloberst von. *Gedanken eines Soldaten*. Leipzig: v. Hase und Koehler Verlag, n.d.

Shaw, Stanford J. *Ottoman Egypt in the Age of the French Revolution*. Cambridge, Mass.: Harvard University Press, 1964.

Shou P'eng-fei 壽鵬飛. "Tung-san-sheng yen-cheng kai-ko i-chien-shu" 東三省鹽政改革意見書 (A program for the reform of the Manchurian salt administration), *Yen-cheng tsa-chih*, No. 3, *chuan-chien* 2:1–18 (March 1913).

Shuster, W. Morgan. *The Strangling of Persia*. New York, 1912.

Sigel, Louis T. "Ch'ing Tibetan Policy (1906–1910)," *Papers on China*, 20:177–201. Harvard University, East Asian Research Center, 1966.

Spector, Stanley. *Li Hung-chang and the Huai Army*. Seattle: University of Washington Press, 1964.

Spence, Jonathan D. *Ts'ao Yin and the K'ang-hsi Emperor: Bondservant and Master*. New Haven and London: Yale University Press, 1966.

Spencer, J. E. "Salt in China," *Geographical Review*, 25.3:353–366 (July 1935).

Ssu-ch'uan yen-fa chih 四川鹽法志 (Treatise on the Szechwan salt laws), comp. Ting Pao-chen 丁寶楨. 40 chüan. Chengtu, 1882.

Stanley, C. John. *Late Ch'ing Finance: Hu Kuang-yung as an Innovator*. Cambridge, Mass.: Harvard University Press, 1961.

Sun E-tu Zen. *Chinese Railways and British Interests, 1898–1911*. New York: King's Crown Press 1954.

——— *Ch'ing Administrative Terms*. Cambridge, Mass.: Harvard University Press, 1961.

Sun I-shuan. "Salt Taxation in China." Ph.D. dissertation, University of Wisconsin, 1953.

Ta Ch'ing li-ch'ao shih-lu 大淸歷朝實錄 (Veritable records of successive reigns of the Ch'ing dynasty). Taiwan: T'ai-wan hua-wen shu-chü, 1963.

T'ang Hsiang-lung 湯象龍. "Min-kuo i-chien kuan-shui tan-pao chih wai-chai" 民國以前關稅擔保之外債 (Prerepublican foreign debt secured on the customs), in Pao Tsun-peng 包遵彭 et al. eds. *Chung-kuo chin-tai-shih lun-ts'ung* 中國近代史論叢 (Collection of essays on modern Chinese history), III, 89–114. Second Series. Taipei: Cheng-chung shu-chü, 1958.

T'an-yen ts'ung-pao 談鹽叢報 (Salt discussion miscellany). Shanghai, 1913–1914.

Tawney, R. H. *Business and Politics under James I*. Cambridge, England: Cambridge University Press, 1958.

Trevor-Roper, H. R. *The Crisis of the Seventeenth Century*. New York and Evanston: Harper and Row, 1968.

Tseng Yang-feng 曾仰豐. "I Ting-en" 憶丁恩 (In memory of Dane), *Yen-yeh t'ung-hsün* 鹽業通訊 (Salt trade bulletin), 77:10–11 (January 1958).

——— *Chung-kuo yen-cheng shih* 中國鹽政史 (History of the Chinese salt administration). Taipei: T'ai-wan shang-wu yin-shu kuan, 1966.

Tu Fang, Lienche. "An Account of the Salt Industry at Tzu-liu-ching: *Tzu-liu-ching chi* by Li Jung," *Isis*, 39:228–234 (1948).

Walker, K. R. "A Chinese Discussion on Planning for Balanced Growth: a Summary of the Views of Ma Yin-ch'u and His Critics," in C. D. Cowan, ed., *The Economic Development of China and Japan*, pp. 160–191. London: George Allen and Unwin Ltd., 1964.

Waugh, Sir Telford. *Turkey, Yesterday, Today and Tomorrow*. London, 1930.

Weber, Max. *The Religion of China*. Glencoe, Ill.: The Free Press, 1951.

Wee, H. van der. *The Growth of the Antwerp Market and the European Economy*. 3 vols. The Hague: Martinus Nijhoff, 1963.

Who's Who in China. 3rd ed. Shanghai, 1926.

Woodward, E. L., and Rohan Butler, eds. *Documents on British Foreign Policy, 1919–1939*. First Series. Vol. VI. *1919*. London: Her Majesty's Stationery Office, 1956.

Wright, Mary C. *The Last Stand of Chinese Conservatism*. Stanford: Stanford University Press, 1962.

Wright, Stanley F. *Hart and the Chinese Customs*. Belfast: Wm. Mullan and Son Ltd., 1950.

Wu Lien-teh. *Plague Fighter: The Autobiography of a Modern Chinese Physician*. Cambridge, England: W. Heffer and Sons Ltd., 1959.

Wu To 吳鐸. "Ch'uan-yen kuan-yün chih shih-mo" 川鹽官運之始末 (A history of the "official transport" of Szechwan salt), *Chung-kuo chin-tai ching-chi-shih yen-chiu chi-k'an* 中國近代經濟史研究集刊 (Studies in modern economic history of China), 3.2:143–261 (1935).

Yang Lien-sheng. "Great Families of Eastern Han," in E-tu Zen Sun and John de Francis, eds., *Chinese Social History*. Washington, D.C.: American Council of Learned Societies, 1956, pp. 103–134.

Yen-cheng tsa-chih 鹽政雜誌 (Salt administration magazine). Peking, 1912–1915.

Young, Arthur N. *China's Wartime Finance and Inflation, 1937–1945.* Cambridge, Mass.: Harvard University Press, 1965.

Yu, George T. *Party Politics in Republican China*. Berkeley and Los Angeles: University of California Press, 1966.

Glossary

An-i-hsien 安邑縣
an-shang 岸商
ao-shang 廒商

Chang Chien 張謇
Chang Hu 張弧
ch'ang 場
ch'ang chü-shen 廠巨紳
ch'ang-kuan 場官
Ch'ang-lu 長蘆
ch'ang-shang 場商
Chao Erh-hsün 趙爾巽
Ch'ao-ch'iao 潮橋
cheng-k'o 正課
chi-an 計岸
chi-ch'u 濟楚
chi-ssu yin 緝私銀
chia-chia yin 加價銀
chia-k'o yin 加課銀
chia-li 加釐
chia-li chü 加釐局
chia-shui 加稅
chiang-fang 江防
chiang-fang ching-fei 江防經費
Chien-wei 犍爲
chien-yen 煎鹽
chih-ying chü 支應局
Chin-ling fang-ying chih-ying chü 金陵防營支應局
Chin-pei 晉北
Ching Pen-po 景本白
chiu-ch'ang cheng-shui 就場征稅
chiu-ch'ang cheng-shui jen ch'i so chih 就場征稅任其所之
chiu-ch'ang cheng-shui tzu-yu fan-mai 就場徵稅自由販賣
chiu-ch'ang chuan-mai 就場專賣
chiu-ch'ang shou-shui 就場收稅
Chiu-ta 久大
Chou Hsüeh-hsi 周學熙
ch'ou-hsiang 籌餉
Ch'u-li yin 楚釐銀
chuan-mai chih 專賣制
Ch'uan 川
Ch'uan-nan 川南
Ch'uan-pei 川北
ch'uan-shui yin 船稅銀
Ch'uan-yen tsung-chü 川鹽總局
Chuang Sung-fu 莊崧甫
ch'ung-yen chih-chao 重鹽執照
chü-fei yin 局費銀

fa-yen chao 發鹽照
Fan Hsü-tung 范旭東
Fan Kao-p'ing 范高平
fang-jen chu-i 放任主義
Feng Kuei-fen 馮桂芬
fu-shui hsi 賦稅系
Fu-shun 富順

hai-fang ching-fei 海防經費
Han-ku 漢沽

hang-shang 行商
Ho-tung 河東
Hsi-pa 西壩
hsiao 梟
Hsiao K'un 蕭堃
hsieh-hsiang yin 協餉銀
hsin-an p'ei-k'uan 新案賠款
Hsiung Hsi-ling 熊希齡
Hsiung K'o-wu 熊克武
Hu-pei ch'iang-p'ao chü 湖北槍砲局
hu-p'iao 護票
Hu-shang 湖商
Hua-ma 花馬
Hua-ting 花定
Huai-chün chuan-yün chü 淮軍轉運局
Huai-li yin 淮釐銀
Huai-nan 淮南
Huai-nan tsung-chü 淮南總局
Huai-pei 淮北
huo-hao yin 火耗銀

kan-she chu-i 干涉主義
kang 綱
ko-sheng yen-yeh hsieh-hui 各省鹽業協會
K'ou-pei 口北
Ku Yen-wu 顧炎武
kuan-shou 官收
kuan-shou, kuan-yün, kuan-mai 官收,官運,官賣
kuan-shou shang-mai 官收商賣
kuan-tu shang-hsiao 官督商銷
kuan-tu shang-pan 官督商辦
kuan-yün 官運
kuan-yün chü 官運局
kuan-yün kuan-hsiao 官運官銷
kuan-yün shang-hsiao 官運商銷
kuei 櫃
kuei-shang 櫃商
kung-fei 公費

Li Wen 李雯
Liang-Che 兩浙
Liang-Huai 兩淮
Liang-chiang chia-li chü 兩江加釐局
lien-ping hsin-hsiang 練兵新餉
lien-ping hsin-hsiang chü 練兵新餉局
Liu Yen 劉晏
Lu-chou 瀘州
lu-hao 滷耗
Lu Jung-t'ing 陸榮廷
lun-ch'uan t'an-fei yin 輪船炭費銀

Ma Yin-ch'u 馬寅初
min-chih, kuan-shou, shang-yün, min-mai 民製官收商運民賣
min-yün min-hsiao 民運民銷

Nei-wu fu 內務府

Pan-p'u 板浦
pao-hsiao yin 報効銀
p'iao 票
p'iao-an 票岸
p'iao-shang 票商
pien-an 邊岸
p'ing-nan-kuei 平南櫃
p'u-fan 鋪販

shai 晒
shai-yen 晒鹽
shang-yün 商運
shih-an 食岸
Shih-erh-wei 十二圩
shou-na 收納
shui-fan 水販
Ssu-ch'uan chüan-sheng ching-shang kung-so 四川全省井商公所
ssu-ma 司碼
Su Wu-shu 蘇五屬
Sung-chiang 松江

Ta-t'ung 大通
T'ang Chi-yao 唐繼堯
T'ang Chiung 唐炯
T'ao Chu 陶澍
Teng Hsiao-k'o 鄧孝可
t'i-chü 提擧
t'ieh-shang-p'ing yü-yin 貼商平餘銀
t'ien-hsia chieh kuan-yen, t'ien-hsia chieh ssu-yen 天下皆官鹽 天下皆私鹽

t'ien-hsia chieh ssu-yen, t'ien-hsia chieh kuan-yen 天下皆私鹽 天下皆官鹽
Ting Pao-chen 丁寶楨
tsa-k'uan 雜欵
Tsai-tse 載澤
ts'ang-tan 艙單
Ts'en Ch'un-hsüan 岑春煊
tso-chao 左照
Tso Shu-chen 左樹珍
tu-hsiao tsung-chü 督銷總局
tu-hsiao fen-chü 督銷分局
tu-pan yen-cheng ch'u 督辦鹽政處
tu-pan yen-cheng ta-ch'en 督辦鹽政大臣
Tzu-liu-ching 自流井
tzu-yu mao-i 自由貿易

Wang Chen-kan 王楨幹
Wang Shao-chi 王紹基

Yen An-lan 晏安瀾
yen-cheng 鹽政
yen-cheng chü 鹽政局
yen-cheng pu 鹽政部
Yen-cheng t'ao-lun hui 鹽政討論會
yen-cheng yüan 鹽政院
yen-cheng yüan yen-cheng ch'eng 鹽政院鹽政丞
yen-cheng yüan yen-cheng ta-ch'en 鹽政院鹽政大臣
yen-fa tao 鹽法道
yen-hao 鹽號
yen-li 鹽釐
yen-wu 鹽務
yen-wu shu 鹽務署
yen-yün shih 鹽運使
yin 引
yin-chieh 引界
yin-chih 引制
yin-shang 引商
yin-ti 引地
ying-yeh hsi 營業系
Yü-yao 餘姚
Yüeh 粵
Yün-ch'eng 運城
yün-shang 運商

Index

Harvard East Asian Series

1. *China's Early Industrialization: Sheng Hsuan-huai (1884–1916) and Mandarin Enterprise.* By Albert Feuerwerker.
2. *Intellectual Trends in the Ch'ing Period.* By Liang Ch'i-ch'ao. Translated by Immanuel C. Y. Hsü.
3. *Reform in Sung China: Wang An-shih (1021–1086) and His New Policies.* By James T. C. Liu.
4. *Studies on the Population of China, 1368–1953.* By Ping-ti Ho.
5. *China's Entrance into the Family of Nations: The Diplomatic Phase, 1858–1880.* By Immanuel C. Y. Hsü.
6. *The May Fourth Movement: Intellectual Revolution in Modern China.* By Chow Tse-tsung.
7. *Ch'ing Administrative Terms: A Translation of the Terminology of the Six Boards with Explanatory Notes.* Translated and edited by E-tu Zen Sun.
8. *Anglo-American Steamship Rivalry in China, 1862–1874.* By Kwang-Ching Liu.
9. *Local Government in China under the Ch'ing.* By T'ung-tsu Ch'ü.
10. *Communist China, 1955–1959: Policy Documents with Analysis.* With a foreword by Robert R. Bowie and John K. Fairbank. (Prepared at Harvard University under the joint auspices of the Center for International Affairs and the East Asian Research Center.)
11. *China and Christianity: The Missionary Movement and the Growth of Chinese Antiforeignism, 1860–1870.* By Paul A. Cohen.
12. *China and the Helping Hand, 1937–1945.* By Arthur N. Young.
13. *Research Guide to the May Fourth Movement: Intellectual Revolution in Modern China, 1915–1924.* By Chow Tse-tsung.
14. *The United States and the Far Eastern Crises of 1933–1938: From the Manchurian Incident through the Initial Stage of the Undeclared Sino-Japanese War.* By Dorothy Borg.
15. *China and the West, 1858–1861: The Origins of the Tsungli Yamen.* By Masataka Banno.

16. *In Search of Wealth and Power: Yen Fu and the West.* By Benjamin Schwartz.
17. *The Origins of Entrepreneurship in Meiji Japan.* By Johannes Hirschmeier, S.V.D.
18. *Commissioner Lin and the Opium War.* By Hsin-pao Chang.
19. *Money and Monetary Policy in China, 1845–1895.* By Frank H. H. King.
20. *China's Wartime Finance and Inflation, 1937–1945.* By Arthur N. Young.
21. *Foreign Investment and Economic Development in China, 1840–1937.* By Chi-ming Hou.
22. *After Imperialism: The Search for a New Order in the Far East, 1921–1931.* By Akira Iriye.
23. *Foundations of Constitutional Government in Modern Japan, 1868–1900.* By George Akita.
24. *Political Thought in Early Meiji Japan, 1868–1889.* By Joseph Pittau, S.J.
25. *China's Struggle for Naval Development, 1839–1895.* By John L. Rawlinson.
26. *The Practice of Buddhism in China, 1900–1950.* By Holmes Welch.
27. *Li Ta-chao and the Origins of Chinese Marxism.* By Maurice Meisner.
28. *Pa Chin and His Writings: Chinese Youth Between the Two Revolutions.* By Olga Lang.
29. *Literary Dissent in Communist China.* By Merle Goldman.
30. *Politics in the Tokugawa Bakufu, 1600–1843.* By Conrad Totman.
31. *Hara Kei in the Politics of Compromise, 1905–1915.* By Tetsuo Najita.
32. *The Chinese World Order: Traditional China's Foreign Relations.* Edited by John K. Fairbank.
33. *The Buddhist Revival in China.* By Holmes Welch.
34. *Traditional Medicine in Modern China: Science, Nationalism, and the Tensions of Cultural Change.* By Ralph C. Croizier.
35. *Party Rivalry and Political Change in Taishō Japan.* By Peter Duus.
36. *The Rhetoric of Empire: American China Policy, 1895–1901.* By Marilyn B. Young.
37. *Radical Nationalist in Japan: Kita Ikki, 1883–1937.* By George M. Wilson.
38. *While China Faced West: American Reformers in Nationalist China, 1928–1937.* By James C. Thomson Jr.
39. *The Failure of Freedom: A Portrait of Modern Japanese Intellectuals.* By Tatsuo Arima.
40. *Asian Ideas of East and West: Tagore and His Critics in Japan, China, and India.* By Stephen N. Hay.
41. *Canton under Communism: Programs and Politics in a Provincial Capital, 1949–1968.* By Ezra F. Vogel.
42. *Ting Wen-chiang: Science and China's New Culture.* By Charlotte Furth.
43. *The Manchurian Frontier in Ch'ing History.* By Robert H. G. Lee.
44. *Motoori Norinaga, 1730–1801.* By Shigeru Matsumoto.
45. *The Comprador in Nineteenth Century China: Bridge between East and West.* By Yen-p'ing Hao.
46. *Hu Shih and the Chinese Renaissance: Liberalism in the Chinese Revolution, 1917–1937.* By Jerome B. Grieder.

47. *The Chinese Peasant Economy: Agricultural Development in Hopei and Shantung, 1890–1949.* By Ramon H. Myers.
48. *Japanese Tradition and Western Law: Emperor, State, and Law in the Thought of Hozumi Yatsuka.* By Richard H. Minear.
49. *Rebellion and Its Enemies in Late Imperial China: Militarization and Social Structure, 1796–1864.* By Philip A. Kuhn.
50. *Early Chinese Revolutionaries: Radical Intellectuals in Shanghai and Chekiang, 1902–1911.* By Mary Backus Rankin.
51. *Communication and Imperial Control in China: Evolution of the Palace Memorial System, 1693–1735.* By Silas H. L. Wu.
52. *Vietnam and the Chinese Model: A Comparative Study of Nguyễn and Ch'ing Civil Government in the First Half of the Nineteenth Century.* By Alexander Barton Woodside.
53. *The Modernization of the Chinese Salt Administration, 1900–1920.* By S. A. M. Adshead.